T0225088

Quantitative Applications in the Social Sciences

A SAGE PUBLICATIONS SERIES

Quantitative Applications in the Social Sciences

A SAGE PUBLICATIONS SERIES

Series/Number 07-167

THE ASSOCIATION GRAPH AND THE
MULTIGRAPH FOR LOGLINEAR MODELS

Harry J. Khamis

Statistical Consulting Center, Wright State University

Los Angeles | London | New Delhi
Singapore | Washington DC

For information:

SAGE Publications, Inc.
2455 Teller Road
Thousand Oaks, California 91320
E-mail: order@sagepub.com

SAGE Publications India Pvt. Ltd.
B 1/I 1 Mohan Cooperative
   Industrial Area
Mathura Road, New Delhi 110 044
India

SAGE Publications Ltd.
1 Oliver's Yard
55 City Road
London EC1Y 1SP
United Kingdom

SAGE Publications Asia-Pacific
   Pte. Ltd.
33 Pekin Street #02-01
Far East Square
Singapore 048763

Library of Congress Cataloging-in-Publication Data

Khamis, Harry.
The association graph and the multigraph for loglinear models / Harry J. Khamis.
    p. cm.—(Quantitative applications in the social sciences; 167)
Includes bibliographical references and index.
ISBN 978-1-4129-7238-3 (pbk.)
    1. Statistics—Graphic methods. 2. Log-linear models. I. Title.

HA31.K45 2011
519.5′35—dc22   2010037414

This book is printed on acid-free paper.

11   12   13   14   15   10   9   8   7   6   5   4   3   2   1

Acquisitions Editor:	Vicki Knight
Associate Editor:	Lauren Habib
Editorial Assistant:	Kalie Koscielak
Production Editor:	Brittany Bauhaus
Copy Editor:	QuADS Prepress (P) Ltd.
Typesetter:	C&M Digitals (P) Ltd.
Proofreader:	Charlotte J. Waisner
Indexer:	Diggs Publication Services, Inc.
Cover Designer:	Candice Harman
Marketing Manager:	Dory Schrader
Permissions Editor:	Adele Hutchinson

CONTENTS

ABOUT THE AUTHOR

Harry J. Khamis is a professor in the Department of Mathematics and Statistics with a joint appointment in the Boonshoft School of Medicine at Wright State University, Dayton, Ohio. He has been Director of the Statistical Consulting Center at Wright State University since 1993. He has been with Wright State University since 1980, except for visiting teaching, research, and consulting positions at Uppsala, Umeå, and Dalarna universities in Sweden. Specializing in statistical methodology, especially categorical response models, goodness-of-fit tests, and the Cox regression model, he has authored or coauthored over 80 publications. In addition to teaching and research, he also consults extensively with researchers at the university as well as clients outside the university. His major external clients include Astra-Arcus Pharmaceuticals, B. F. Goodrich, Cancer Prevention Institute, Carnation Co., Center for Election Integrity, Clinical Research Consultants, Community Blood Center, Genentech Inc., Isolab Inc., Kunesh Eye Surgery Center, Mandal Diabetes Research Institute, Nestlé's, Pharmacia-Upjohn Pharmaceuticals, and Sifo Marketing Research. He received his BS degree in mathematics at Santa Clara University, and he received his MS degree in mathematics (1976) and his PhD degree in statistics (1980) at Virginia Tech.

SERIES EDITOR'S INTRODUCTION

The development and popularization of loglinear models (LLMs) in the late 1960s and early 1970s by Leo Goodman and others, summarized in the landmark 1975 text *Discrete Multivariate Analysis* by Bishop, Fienberg, and Holland, revolutionized the analysis of multi-way contingency tables. Courses on categorical data analysis, including but not restricted to LLMs, are now a staple of graduate education in the social sciences. Harry Khamis's monograph *The Association Graph and the Multigraph for Loglinear Models* shows how the mathematical resources of graph theory can be mobilized to understand the structure of associations implied by a complex LLM.

Khamis reviews the patterns of association in two- and higher-way contingency tables, along with LLMs for these tables. After introducing some key concepts from graph theory, he proceeds to apply these ideas to two graphical representations of LLMs: the *association graph* and the *generator multigraph*. In the more straightforward and widely known association graph, the *vertices* (points) represent the variables in a contingency table, and the *edges* (lines) correspond to association terms in the model. In the multigraph, in contrast, the vertices represent the high-order association terms (*generators*) of a hierarchical LLM, and the edges correspond to variables that are shared by pairs of generators.

Deploying a wealth of carefully selected examples and employing lucid explanations, Khamis shows how both graphical representations of LLMs illuminate the structure of the models, revealing characteristics such as conditional independence (e.g., where variables A and B are independent within categories of variable C), decomposability (e.g., where the joint cell probabilities for A, B, and C implied by the model are the products of the marginal probabilities for AB and for C), and collapsibility (e.g., where the marginal association between A and B is the same as the association between A and B within categories of C). Taken together, the graphical methods described in this monograph facilitate the formulation and interpretation of complex LLMs for high-dimensional tables. The monograph should prove valuable both to students who wish to develop a more sophisticated understanding of the structure of LLMs and to researchers who apply these models in research.

Editor's note: This monograph was begun under the direction of the previous series editor, Tim Futing Liao.

—*John Fox*

Series Editor

Il faut toujours prendre la mesure des choses avant de décider.

CHAPTER 1. INTRODUCTION

While sophisticated statistical methodologies for continuous outcome variables began in the 1880s with Francis Galton's work, and later with the work of, among others, R. A. Fisher and Karl Pearson, substantial breakthroughs were not made for categorical response models until the 1960s despite influential articles by Karl Pearson and G. Udny Yule as far back as 1900. Methodologies for handling categorical response models have exploded over the past 50 years. A very brief historical outline of the methodologies used to analyze multidimensional contingency tables might include the following developments:

- G. Udny Yule's work in the early 1900s established the odds ratio as a key measure of association between attributes; more than 60 years later, odds ratios would become the building blocks of loglinear models (LLMs; see Chapter 3 for a review of LLMs).
- M. S. Bartlett (1935) presented one of the first analyses of a three-way contingency table, including analysis of first- and second-order interaction effects.
- Since the 1960s, research work on LLMs of multi-way contingency tables blossomed. M. W. Birch (1963) solved the maximum likelihood estimation problem for the LLM of a three-way contingency table. Grizzle, Starmer, and Koch (1969) presented weighted least squares methods for categorical data models. Bishop, Fienberg, and Holland (1975) wrote the first definitive textbook on the analysis of multi-way contingency tables. Since then, many texts have been written dealing with categorical response models. Extensive contributions to the categorical data literature have been made by individuals such as C. C. Clogg, D. R. Cox, J. N. Darroch, I. J. Good, L. A. Goodman, S. J. Haberman, P. F. Lazarsfeld, R. L. Plackett, N. Wermuth, and many others.
- Darroch, Lauritzen, and Speed (1980) wrote a landmark article introducing a way of representing LLMs using mathematical graphs and using the properties of mathematical graphs to analyze and interpret contingency table data. Since 1980, much work has been done on these so-called first-order interaction graphs, also called *association graphs* by Agresti (2002), especially by researchers such as D. R. Cox, D. Edwards, S. L. Lauritzen, and N. Wermuth. This marriage between mathematical graph theory and LLM theory provides a

revolutionary way of analyzing and interpreting the complex structural relationships among categorical variables. Two good resources in this regard are the books by Edwards (1995) and Lauritzen (1996).

Two developments in recent decades in the psychological, sociological, political science, educational, and other research environments have led to a need for more sophisticated tools with which to analyze and interpret large, complex LLMs.

1. First, data collection and data entry practices in recent decades have enabled researchers to acquire extremely large data sets, especially since the full power of the computer and electronic scanning devices became available. In studies involving categorical variables, this results in large multidimensional contingency tables with large sample sizes. In fact, one need only access the Internet to find massive data sets available for use by researchers. Many of these data sets are provided by federal or state agencies for public use. An excellent example of this is the Statistiska Centralbyrån (Statistics Sweden), which makes available data for research purposes (see www.scb.se). Another example is the Bureau of Labor Statistics (http://bls.gov/nls/nlsy79.htm).

2. Second, the notion of "sparseness" in contingency tables has changed. In the early years of contingency table analysis, statisticians advised that chi-squared analysis be employed only if the expected frequency for each cell is at least 5 (Fisher, 1925). However, there is evidence that such analyses work well with sparser tables. For instance, Cochran (1954) states that chi-squared analysis is valid if the expected cell frequency is at least 1 for all cells and not more than 20% of the cells have expected frequencies less than 5. More recently, a review and empirical analysis by Roscoe and Byars (1971) provided some additional useful guidelines. They recommend that the average expected frequency per cell be at least 6 when testing at the .05 level of significance. If conditional tests are used, considerably sparser tables are allowable (see Whittaker, 1990, chap. 9, for an example). Wickens (1989, p. 30), citing several authors, suggests that the total sample size should be at least four or five times the number of cells in a multiway table when testing the goodness of fit of a model. Lawal and Upton (1984) discuss modifications of the chi-squared test to allow the average expected cell frequency to be as small as 0.5.

These two developments expand the applicability of contingency table analysis methods, enabling the researcher to include ever larger numbers of variables in the model. This is especially useful to the practitioner given the importance of including all potential confounders in a given analysis so as to adjust for their effects.

Thus, the LLM can be fit to data from large, complex, multidimensional contingency tables using the Pearson chi-squared or likelihood ratio chi-squared statistics effectively, even if the average expected cell frequencies are quite low. Once a best-fitting LLM is identified, we wish to analyze and interpret it, effectively determining what the data say about the behavior of the variables according to the selected model. Because the structural relationships among all the factors of a large multidimensional contingency table can be complex, especially for those who are not expert in handling LLMs, researchers have made efforts to develop methodologies and strategies to better analyze and interpret the LLMs of such tables. In this regard, the first-order interaction graph has become very useful. More recently, an alternative graphical representation of LLMs was introduced—namely, the *generator multigraph*, or simply *multigraph* (McKee & Khamis, 1996; see also Khamis, 1996, 2005).

Generally, the analysis of the relationships among a set of categorical variables using LLMs can be done in two stages:

1. Identifying the model that fits the data "best"
2. Analyzing and interpreting the resulting best-fitting model

The first of these two stages is, to be sure, a nontrivial task but one for which there is an extensive literature leading to a variety of model-fitting procedures and strategies, including conventional LLM-fitting procedures as well as more specialized procedures such as conditional tests, bootstrap procedures, Bayesian approaches, and so on. Statistical methodologies and software are available to find the best-fitting LLM for a given contingency table, for example, stepwise procedures (Goodman, 1971a) and the two-step procedure (Benedetti & Brown, 1978; Brown, 1976). See also Agresti (2002, chap. 9), Wickens (1989, chap. 5), and Lawal (2003, chap. 7). This book does not cover the procedures for finding the best-fitting LLM. Rather, the reader is referred to the literature cited above. In each example in this book, the best-fitting LLM (or at least an LLM that fits well) for a given set of data will be provided, with little or no discussion.

This book concerns the latter of the two stages given above. Once a best-fitting LLM is obtained, it is essential that it be analyzed and interpreted correctly and thoroughly. In this book, "analyzing" a given LLM refers to finding the important properties of the model; "interpreting" a given LLM refers to identifying all the relationships among the variables and translating those relationships into conclusions about the data. Using the tools available in mathematical graph theory makes this aspect of the overall statistical analysis reliable, organized, comprehensive, and easy. The first graphical procedure, the association graph, is already included in many standard categorical data textbooks (e.g., Agresti, 2002; Andersen, 1997; Wickens, 1989). The second procedure, the multigraph, is relatively new and has not appeared in textbooks.

It is assumed that the reader is familiar with the application of the LLM to data obtained from a multidimensional contingency table generated by one of the standard sampling designs (see Subsection 3.4.3), and with the procedures for selection of a best-fitting model. This book focuses on the analysis and interpretation of the resulting best-fitting LLM. Copious real-data examples from the psychology, political science, and sociology literature are provided for illustration purposes. Many of the real-world data examples presented in this book are taken (with client permission) from projects received by the Statistical Consulting Center at Wright State University.

This book will serve generally as a "how to" guide, focusing on the practical application of the association graph and the multigraph to a best-fitting LLM for the purpose of thoroughly and reliably analyzing and interpreting it. The reader will be referred to the appropriate literature for the theory underlying the methodologies presented, including theorems, proofs, derivations, and calculation methods. On mastering the material in this book, the reader will be able to interpret a very complex LLM by

1. identifying the important properties of the model, thereby leading to a deeper understanding of the model;

2. explaining the relationships among the factors in a clear and understandable manner; and

3. determining ways in which the contingency table can be simplified (e.g., by the use of collapsibility conditions).

Finally, these goals can be accomplished in a facile manner through the use of the association graph and/or the multigraph. And while the search for a best-fitting model requires computer software and possibly complex model selection strategies and techniques, once the best-fitting model has been found, it can be analyzed and interpreted using graphical procedures, without the need for any computer software, complex derivations, or heavy calculations.

For most LLMs involving up to four variables (or perhaps five variables depending on the complexity of the model), the relationships among the variables can be determined merely by scrutinizing the LLM itself or the generating class (see Chapter 3). However, for more complex LLMs, sorting out all the information contained in the model can be extremely challenging, even for the most experienced LLM analyst. It is for these complex LLMs based on large multi-way contingency tables that the procedures in this book are most useful.

As a motivational example, consider 10 categorical variables coded 0, 1, 2, . . . , 9. It is desired to know the relationships among these 10 variables. Suppose the generating class (also called the minimal sufficient configuration) for the best-fitting LLM of the 10-way contingency table is [67][013][125][178][1347][1457][1479]. Which factors are independent of which others? Which factors are conditionally independent of which others? Can you be sure that you've identified *all* the independencies and conditional independencies? Which factors can be collapsed over without changing the associations among the other factors? Can you be sure that you've identified *all* the associations that are preserved under the collapsing? What are the important properties of this model? It is not easy to answer these questions thoroughly and reliably from the generating class alone, even for LLM experts. The procedures presented in this book enable the researcher to answer these questions thoroughly and reliably in a clear, comprehensive, organized, step-by-step manner that requires no statistical software or heavy calculations. The researcher will thus be able to understand clearly how the factors interrelate and, most important, how to interpret the data accurately and thoroughly. This particular 10-variable model serves as an illustrative example in the following chapters.

This book is organized as follows. Structures of association are defined and discussed in Chapter 2, LLMs and their properties are reviewed in Chapter 3, the association graph is presented and discussed in Chapters 4 and 5, and the multigraph is introduced and discussed in Chapters 6 and 7. Conclusions and additional illustrative real-data examples are provided in Chapter 8.

CHAPTER 2. STRUCTURES OF ASSOCIATION

In this book, *structures of association* or *structural associations* refer to the independencies and conditional independencies (and the absence thereof) among the factors in a multidimensional contingency table. One of the most important goals in the analysis of contingency table data is to accurately and thoroughly identify the structural associations among the categorical variables. When only two categorical variables are involved, then the statistical significance of the association between the two variables can be assessed by using the well-known chi-squared test. When three or more variables are involved, this task becomes more complicated. In this chapter, we discuss the structures of association for two- and three-way contingency tables using the odds ratio as the principal measure of association.

2.1 Statistical Independence for Discrete Variables

For categorical data, measures of association are used to quantify the degree of relationship (or association) between two variables. Consider a two-way contingency table having I rows and J columns, corresponding to the two factors X and Y, respectively. The two discrete variables, X and Y, are statistically *independent* if their joint probability can be factored into the product of the marginal probabilities; notationally,

$$\pi_{ij} = \pi_{i+}\pi_{+j}, \qquad i = 1, 2, \ldots, I \text{ and } j = 1, 2, \ldots, J, \qquad (2.1)$$

where π_{ij} is the probability that a subject falls in the ith level of X and the jth level of Y, π_{i+} is the marginal probability that a subject falls in the ith level of X, and π_{+j} is the marginal probability that a subject falls in the jth level of Y. Using Goodman's (1970) notation, we write "$[X \otimes Y]$" to represent the statistical independence between X and Y:

$[X \otimes Y]$ *if and only if* X *and* Y are statistically independent.

To the degree that π_{ij} deviates from $\pi_{i+}\pi_{+j}$, the two factors are statistically *dependent* or *associated*. There is a wide variety of association measures for categorical variables, depending on the measurement levels of the variables, whether one variable is antecedent to the other, and so on. For a review of such measures see Goodman and Kruskal (1979) or Khamis (2004).

6

2.2 The Odds Ratio: Two-Way Tables

When working with loglinear models (LLMs), the association measure of choice is the *odds ratio* (defined below), since LLM parameters are functions of odds ratios. Consequently, we will limit our discussion to the odds ratio as the principal measure of association for the remainder of this chapter.

For a 2 × 2 contingency table, the *odds* of a subject being in the first column rather than in the second column given that it is in the first row is π_{11}/π_{12}, and the odds that a subject is in the first column rather than in the second column given that it is in the second row is π_{21}/π_{22}. (In this context, "subject" is the thing being measured and cross-classified; in other contexts, it might just as well be a patient, a mouse, a school, etc.) Then the *odds ratio,* denoted by α, is the ratio of the two odds:

$$\alpha = \frac{\pi_{11}/\pi_{12}}{\pi_{21}/\pi_{22}}, \qquad \text{or simply } \alpha = \frac{\pi_{11}\pi_{22}}{\pi_{12}\pi_{21}}. \tag{2.2}$$

When the odds that an item is in the first column rather than in the second column is the same for both rows, then $\alpha = 1$; this occurs when X and Y are statistically independent. Alternatively, one can replace π_{ij} by $\pi_{i+}\pi_{+j}$ (see Equation 2.1) in Equation 2.2 to prove that $\alpha = 1$ when X and Y are independent. The maximum likelihood estimator of α is $\hat{\alpha} = n_{11}n_{22}/n_{12}n_{21}$, where n_{ij} is the observed cell frequency for the ith row and jth column, with $i = 1, 2$ and $j = 1, 2$.

The odds ratio is defined on the interval $[0, \infty)$ and takes the value 1.0 when the two variables, say X and Y, are independent. Because of its advantageous statistical properties, the natural logarithm ("log") of the estimated odds ratio is typically used in the analysis of categorical data. (*Note:* log(0) is treated as $-\infty$.) Then, statistical independence corresponds to $\log(\alpha) = 0$. Notationally, we would write

$$[X \otimes Y] \ \textit{if and only if } \log(\alpha) = 0.$$

For a large enough sample size and for standard sampling designs (see Subsection 3.4.3), $\log(\hat{\alpha})$ is asymptotically normal and has an estimated standard error:

$$\hat{\sigma}_{\log(\hat{\alpha})} = \sqrt{\frac{1}{n_{11}} + \frac{1}{n_{12}} + \frac{1}{n_{21}} + \frac{1}{n_{22}}}.$$

Example 2.2.1. Table 2.1 presents data from a 1992 survey executed collaboratively by the Wright State University Boonshoft School of Medicine and the United Health Services in Dayton, Ohio. In this survey, 2,102 nonurban Caucasian high school seniors were asked, among other things, if they had ever used cigarettes. It is of interest to know whether there is an association between cigarette use and gender in this population.

Table 2.1. Cross-Classification of 2,102 Nonurban Caucasian High School Seniors According to Cigarette Use and Gender

	Cigarette Use?	
Gender	Yes	No
Male	699	363
Female	691	349

SOURCE: Wright State University Boonshoft School of Medicine and United Health Services 1992 survey, Dayton, Ohio. Data were kindly provided by Dr. Russell Falk.

NOTE: See Example 2.2.1.

For these data,

$$\hat{\alpha} = \frac{(699)(349)}{(691)(363)} = 0.9726, \quad \log(\hat{\alpha}) = -0.0278, \quad \text{and} \quad \hat{\sigma}_{\log(\hat{\alpha})} = 0.0922.$$

The estimated odds of having smoked cigarettes is 2.74% lower for males than for females (1.9256 compared with 1.9799). Equivalently, one may say that the odds of having smoked rather than not smoked is 1.0282 times larger among females than among males. The 95% confidence interval for $\log(\alpha)$ is $-0.0278 \pm (1.96)(0.0922)$, leading to the interval $[-0.2085, 0.1529]$. To get the 95% confidence interval for α, we simply exponentiate, thereby obtaining the interval $[0.812, 1.165]$. Since this interval includes the value 1.0, we do not have strong evidence in these data to conclude that gender is associated with cigarette use in this population.

For $I \times J$ contingency tables, the measure of association between X and Y consists of a set of $(I - 1)(J - 1)$ odds ratios: $\alpha_{ij} = \pi_{ij}\pi_{IJ}/\pi_{Ij}\pi_{iJ}$ for $i = 1, 2, \ldots, I - 1$ and $j = 1, 2, \ldots, J - 1$. Note that the (I, J) cell is an "anchor" for every α_{ij} in this set. This set of odds ratios is not unique; there are other choices of $(I - 1)(J - 1)$ odds ratios that can be used to measure the association between X and Y (see Agresti, 2002, sec. 2.4). The maximum likelihood estimators, $\hat{\alpha}_{ij}$, are obtained by replacing π_{ij} by n_{ij}, as was done for the 2×2 case above. To illustrate, consider the 2×3 contingency table with cell probabilities π_{ij} below:

		Y	
	1	2	3
X 1	π_{11}	π_{12}	π_{13}
2	π_{21}	π_{22}	π_{23}

The association between X and Y is measured by the set of two odds ratios $\pi_{11}\pi_{23}/\pi_{21}\pi_{13}$ and $\pi_{12}\pi_{23}/\pi_{22}\pi_{13}$. Then, X and Y are independent if and only if both odds ratios are equal to 1. That is, $\pi_{ij} = \pi_{i+}\pi_{+j}$ for $i = 1, 2$ and $j = 1, 2, 3$ if and only if $\pi_{ij}\pi_{IJ}/\pi_{Ij}\pi_{iJ} = 1$ for $i = 1$ and $j = 1, 2$. The test for

$$H_0 : \quad \left[\frac{\pi_{11}\pi_{23}}{\pi_{21}\pi_{13}}, \frac{\pi_{12}\pi_{23}}{\pi_{22}\pi_{13}}\right] = [1, 1]$$

can be carried out with the standard chi-squared test for independence.

For more discussion of odds ratios, see Agresti (2002), Rudas (1998), and Fleiss, Levin, and Paik (2003).

2.3 The Odds Ratio: Three-Way Tables

Consider three categorical variables, X (row variable), Y (column variable), and Z (layer variable). It might seem that to assess the structural associations among these three variables, one need only test the associations among all three variables pairwise using a series of simple chi-squared tests: X versus Y, X versus Z, and Y versus Z. However, this is not the best strategy because (a) it ignores the inflation of the Type I error rate for multiple testing, (b) it does not take into account the interaction effects among the variables, and (c) it does not use all the data to full advantage.

When dealing with two categorical variables, there are only two kinds of structural association models to contend with: (1) X and Y are independent or (2) X and Y are dependent. With three categorical variables, there are five kinds of structural association models that are possible. Each of these is described below.

2.3.1. Mutual Independence

All three variables are *mutually independent*. Using Goodman's (1970) notation, we write $[X \otimes Y \otimes Z]$. Using an extension of the probability notation used above (see Equation 2.1), we would have

$$\pi_{ijk} = \pi_{i++}\pi_{+j+}\pi_{++k}, \qquad i = 1, 2, \ldots, I; j = 1, 2, \ldots, J;$$
$$\text{and } k = 1, 2, \ldots, K.$$

In this notation, π_{ijk} represents the probability that a subject falls in the ith row (X), jth column (Y), and kth layer (Z). The notation π_{i++} represents the probabilities corresponding to the levels of X obtained by *collapsing* (or adding) over the levels of Y and Z: $\pi_{i++} = \sum_{j,k} \pi_{ijk}$. This is called the X-*marginal* table (and similarly for π_{+j+} [Y marginal table] and π_{++k} [Z marginal table]). Under mutual independence, the odds ratio (or set of odds ratios) between any two variables is equal to 1.0: $\alpha_{XY} = \alpha_{XZ} = \alpha_{YZ} = 1.0$. For $I > 2$ and/or $J > 2$ and/or $K > 2$, the notation will be relaxed so that, for example, α_{XY} represents a set of $(I - 1)(J - 1)$ odds ratios, and similarly for α_{XZ} and α_{YZ}.

2.3.2. Joint Independence

One of the three variables is independent of the other two. For instance, X is *jointly independent* of Y and Z: $[X \otimes Y, Z]$. The probabilistic model for this case is

$$\pi_{ijk} = \pi_{i++}\pi_{+jk}, \qquad i = 1, 2, \ldots, I; j = 1, 2, \ldots, J;$$
$$\text{and } k = 1, 2, \ldots, K.$$

Note that while $[X \otimes Y]$ and $[X \otimes Z]$, it is not true that $[Y \otimes Z]$. In this case, we have $\alpha_{XY} = \alpha_{XZ} = 1.0$, but $\alpha_{YZ} \neq 1.0$. Finally, there are three ways in which this model can occur: $[X \otimes Y, Z]$, $[Y \otimes X, Z]$, and $[Z \otimes X, Y]$.

2.3.3. Conditional Independence

Two of the variables are independent at each level of the third variable. For instance, X and Y are independent for each level of Z, or *conditional* on the level of Z, X and Y are independent. Notationally, we write $[X \otimes Y|Z]$. The probabilistic model is

$$\pi_{ijk} = \frac{\pi_{i+k}\pi_{+jk}}{\pi_{++k}}, \qquad i = 1, 2, \ldots, I; j = 1, 2, \ldots, J;$$
$$\text{and } k = 1, 2, \ldots, K.$$

Consider the kth layer, where $Z = k$ (this is called the *partial* table at $Z = k$). For this two-way partial table, X and Y are independent. In this case, the odds ratios for X and Y in the partial table corresponding to each level of Z are 1.0: $\alpha_{(XY|Z)} = 1.0$ for $Z = 1, 2, \ldots, K$. Note that while X and Y are independent for each layer, it is not necessarily true that X and Y are independent in the marginal table (see Section 3.3 and Chapter 5 for further discussion of this fact). Finally, there are three ways in which the conditional independence model can occur: $[X \otimes Y|Z]$, $[X \otimes Z|Y]$, and $[Y \otimes Z|X]$.

2.3.4. Homogeneous Association

Any two of the variables are associated, and the association is the same for each level of the third variable. So (1) the X-Y association is homogeneous across layers, (2) the X-Z association is homogeneous across columns, and (3) the Y-Z association is homogeneous across rows. Notationally,

$$1.\ \alpha_{(XY|Z=1)} = \alpha_{(XY|Z=2)} = \cdots = \alpha_{(XY|Z=K)},$$

$$2.\ \alpha_{(XZ|Y=1)} = \alpha_{(XZ|Y=2)} = \cdots = \alpha_{(XZ|Y=J)},$$

$$3.\ \alpha_{(YZ|X=1)} = \alpha_{(YZ|X=2)} = \cdots = \alpha_{(YZ|X=I)}.$$

Since there is no independence between any of the three variables, there is no "$\otimes$" notation to represent this model. Also, the probabilistic model cannot be written in closed form in this case because π_{ijk} cannot be factored into marginal probabilities such as π_{i++}, π_{ij+}, and so on. Models in which the joint probability, π_{ijk}, can be factored in terms of marginal probabilities are called *decomposable* models; they are also called *direct* models, models of *Markov type*, and *multiplicative* models (not to be confused with the use of the term *multiplicative* when dealing with the distinction between additive and multiplicative statistical models). When the joint probability cannot be factored into marginal probabilities in closed form, the model is called a *nondecomposable* model.

Decomposable models were introduced by Goodman (1970, 1971b) and further developed in Haberman (1974) and Andersen (1974). The mutual independence, joint independence, and conditional independence models are all decomposable models; the homogeneous association model is a non-decomposable model. As we will see later, decomposable models have certain advantageous properties.

The X-Y association for each level of Z is called the *conditional* association (also called the *partial* association) between X and Y. The X-Y association in the table obtained by adding over levels of Z is called the *marginal* association between X and Y. As we will see, partial association is not necessarily the same as marginal association (Section 3.3 and Chapter 5).

2.3.5. Saturated Model

Any two of the variables are associated, and the association is not the same for each level of the third variable. So (a) the X-Y association is not the same across all layers, (b) the X-Z association is not the same across all columns, and (c) the *Y-Z* association is not the same across all rows. Notationally, $\alpha_{(XY|Z=k)}$ is not the same for all $k = 1, 2, \ldots, K$; $\alpha_{(XZ|Y=j)}$ is not the same for all $j = 1, 2, \ldots, J$; and $\alpha_{(YZ|X=i)}$ is not the same for all $i = 1, 2, \ldots, I$.

It must be emphasized that the five types of structural associations described above for a three-way contingency table are based on odds ratios as the measures of association. If some other measure of association, some kind of correlation, or percent differences are used to measure association, then these definitions would be different.

2.4 Model Fitting: Three-Way Tables

For a given observed three-way contingency table, one may now compare the observed cell frequencies with the maximum likelihood estimators of the expected cell frequencies, where the estimated expected frequencies are obtained from the probability formulas above for a given LLM. For example, to fit the mutual independence model, calculate the estimated expected cell frequencies from (see Subsection 2.3.1)

$$n\hat{\pi}_{ijk} = n\hat{\pi}_{i++}\hat{\pi}_{+j+}\hat{\pi}_{++k} = n\frac{n_{i++}n_{+j+}n_{++k}}{n^3} = \frac{n_{i++}n_{+j+}n_{++k}}{n^2}. \quad (2.3)$$

Here, $n = \sum_{i,j,k} n_{ijk}$ is the total sample size. To assess the plausibility of the mutual independence model, compare the observed cell frequencies $O = n_{ijk}$ with the estimated expected cell frequencies $E = \frac{n_{i++}n_{+j+}n_{++k}}{n^2}$ using, say, the chi-squared statistic $\chi^2 = \sum_{\text{all cells}} (O - E)^2/E$, and compare this value with, say, the 95th percentile of the $IJK - I - J - K + 2$ degree-of-freedom (df) chi-squared distribution. (See the LLM review in Chapter 3 for calculation of the df for LLMs.) If χ^2 exceeds this percentile (or, equivalently, the P value falls below .05), then the mutual independence model is strongly contradicted by the data.

Example 2.4.1. For the data of Table 2.1, involving cigarette use by high school seniors in Dayton, Ohio, a third variable of interest is race. The pertinent $2 \times 2 \times 2$ table is given in Table 2.2. It is of interest to know whether these three variables are mutually independent. Using $X =$ Gender, $Y =$ Cigarette Use, and $Z =$ Race, we wish to test that $[X \otimes Y \otimes Z]$, so the estimated expected cell frequencies are $n_{i++}n_{+j+}n_{++k}/n^2$ (see Equation 2.3 above). For this mutual independence model, the dfs are $IJK - I - J - K + 2 = 4$. The resulting χ^2 value is 3.36, compared to the 4-df chi-squared distribution, resulting in $P = .4995$. Mutual independence among the factors Gender, Cigarette Use, and Race is not contradicted by the data.

Table 2.2. Cross-Classification of 2,276 Nonurban High School Seniors According to Cigarette Use, Gender, and Ethnic Background

		Cigarette Use?	
Race	*Gender*	*Yes*	*No*
Caucasian	Male	699	363
	Female	691	349
Other	Male	58	36
	Female	47	33

SOURCE: Wright State University Boonshoft School of Medicine and United Health Services 1992 survey, Dayton, Ohio.

NOTE: See Example 2.4.1.

2.5 Multi-Way Tables

The analysis and interpretation of a given best-fitting LLM becomes increasingly complex the greater the number of factors in the contingency table. For instance, with 4 variables there are 12 types of structural association models that can occur, involving mutual independence, joint independencies, conditional independencies, and high-order interaction effects. With 10 variables, this increases to 1,014 types of structural association models. Hence, we need (a) a sophisticated model that helps characterize all the possible types of structural associations and sorts them out, (b) an analysis method that identifies which model fits the data best, and (c) a strategy to help interpret and analyze the resulting best-fitting model.

In answer to (a), the LLM is the model of choice for high-dimensional contingency tables. In answer to (b), many model-fitting strategies have been developed for the LLM, as mentioned in Chapter 1. The remaining chapters of this book are directed toward answering the need mentioned in (c). That is, we assume that a best-fitting LLM has been selected, and we wish to analyze and interpret it. We begin with a brief review of the LLM in the next chapter.

CHAPTER 3. LOGLINEAR MODEL REVIEW

Loglinear models (LLMs) are used to analyze data from a multi-way contingency table. There are several good books to which the reader can refer, including Agresti (2002), Bishop et al. (1975), Christensen (1990), Knoke and Burke (1980), and Wickens (1989). In this chapter, a brief review of the LLM will be given, starting with the two-way table and proceeding to the multi-way table.

3.1 Two-Way Contingency Table

Consider the $I \times J$ contingency table for the two variables X (row) and Y (column). We have seen that X and Y are statistically independent if

$$\pi_{ij} = \pi_{i+}\pi_{+j}, \qquad i = 1, 2 \ldots, I \text{ and } j = 1, 2, \ldots, J.$$

Then,

$$n\pi_{ij} = n\pi_{i+}\pi_{+j}, \qquad i = 1, 2, \ldots, I \text{ and } j = 1, 2, \ldots, J.$$

Let $\mu_{ij} = n\pi_{ij}$ represent the expected cell frequency in the ith row and jth column. Under the model of independence,

$$\mu_{ij} = \frac{\mu_{i+}\mu_{+j}}{n},$$

where $\mu_{i+} = n\pi_{i+}$ represents the marginal expected frequency for the ith row and $\mu_{+j} = n\pi_{+j}$ represents the marginal expected frequency for the jth column. Since additive models are often easier to work with than multiplicative models, we take the natural logarithm:

$$\log(\mu_{ij}) = -\log(n) + \log(\mu_{i+}) + \log(\mu_{+j}).$$

So the expected cell frequency is additive in the logarithmic scale, hence the name *loglinear model* (LLM). This model states that the logarithm of the expected cell frequency is equal to a term involving only the total sample size, $-\log(n)$; a term relying only on the row, $\log(\mu_{i+})$; and a term relying only on the column, $\log(\mu_{+j})$. We denote these terms by λ, λ_i^X, and λ_j^Y, respectively. For identifiability purposes, we must put constraints on these LLM parameters. Though there are different choices about what kinds of constraints to impose, we will use the so-called zero-sum constraints: $\sum_i \lambda_i^X = \sum_j \lambda_j^Y = 0$. Now the complete LLM for independence between X and Y can be written:

$$\log(\mu_{ij}) = \lambda + \lambda_i^X + \lambda_j^Y, \qquad (3.1)$$

where

$$\sum_i \lambda_i^X = \sum_j \lambda_j^Y = 0.$$

It can be shown that λ_i^X and λ_j^Y are functions of $\log(\mu_{ij})$, representing the ith row effect and the jth column effect, respectively. That is,

$$\lambda_i^X = \frac{1}{J}\sum_j \log(\mu_{ij}) - \frac{1}{IJ}\sum_{i,j} \log(\mu_{ij}), \qquad i = 1, 2, \ldots, I,$$

and similarly for λ_j^Y.

For the LLM of dependence, an additional term called the *first-order interaction* between X and Y, denoted by λ_{ij}^{XY}, is added to the model. Along with the zero-sum constraints, the model of dependence between X and Y (also called the *saturated* model) becomes

$$\log(\mu_{ij}) = \lambda + \lambda_i^X + \lambda_j^Y + \lambda_{ij}^{XY}, \qquad (3.2)$$

where

$$\sum_i \lambda_i^X = \sum_j \lambda_j^Y = \sum_i \lambda_{ij}^{XY} = \sum_j \lambda_{ij}^{XY} = \sum_{i,j} \lambda_{ij}^{XY} = 0.$$

Note that because of the constraints, there are only $I - 1$ independent λ_i^X terms, $J - 1$ independent λ_j^Y terms, and $(I - 1)(J - 1)$ independent λ_{ij}^{XY} terms. For the 2×2 contingency table, it can be shown that $\lambda_{11}^{XY} = \log(\alpha)/4$. Thus, the first-order interaction term measures the degree of dependence between X and Y as a function of the odds ratio, α. Note that if $\alpha = 1$ (corresponding to $[X \otimes Y]$), then $\lambda_{11}^{XY} = 0$, hence $\lambda_{12}^{XY} = \lambda_{21}^{XY} = \lambda_{22}^{XY} = 0$. Thus, the LLM of independence (Equation 3.1) obtains.

The degrees of freedom (df) for a given LLM can be calculated in one of two equivalent ways:

1. The total number of cells in the table minus the number of independent λ-terms in the LLM

2. The number of independent λ-terms set to 0 in the saturated model to obtain the LLM

So the *df* calculation for the LLM of independence in the $I \times J$ table would be

1. $IJ - [1 + (I - 1) + (J - 1)] = (I - 1)(J - 1)$ or

2. $\lambda_{ij}^{XY} = 0$ for all i and j in Equation 3.2, to obtain Equation 3.1, hence $(I - 1)(J - 1)$ independent λ_{ij}^{XY} -terms are set to 0.

The *df* calculation for the saturated model is

1. $IJ - [1 + (I - 1) + (J - 1) + (I - 1)(J - 1)] = 0$ or

2. no λ-terms are set to 0 in the saturated model (Equation 3.2), hence $df = 0$.

The notations that will be used for the independence and saturated models in the two-way table are $[X][Y]$ and $[XY]$, respectively. (This notation is valid only for the so-called hierarchical LLMs [see Subsection 3.4.1].) That is, $[X][Y]$ is shorthand for the LLM in Equation 3.1, and $[XY]$ is shorthand for the LLM in Equation 3.2. This notation is referred to as the *generating class* for the LLM; it is also called the *minimal sufficient configuration*. For the independence model, $[X]$ and $[Y]$ are called the *generators* of the LLM; for the saturated model, $[XY]$ is the generator. Note the correspondence between the generators and the λ-terms in the corresponding LLM: In the model of independence (Equation 3.1), the row (X) and column (Y) λ-terms (λ_i^X and λ_j^Y, respectively) appear separately, hence the notation $[X][Y]$; in the saturated model (Equation 3.2), the λ-term representing the combination of row and column, λ_{ij}^{XY}, appears, hence the notation $[XY]$. It can be shown that for the generating class $[X][Y]$, the minimal sufficient statistics for the LLM parameters are the observed row and column marginal totals, $\{n_{i+}\}_{i=1,2,\ldots,I}$ and $\{n_{+j}\}_{j=1,2,\ldots,J}$, respectively. For $[XY]$, the minimal sufficient statistics are the observed cell frequencies, $\{n_{ij}\}_{i=1,2,\ldots,I;j=1,2,\ldots,J}$. Hence, the term *minimal sufficient configuration* is used for the notation. Note the notational correspondence between the minimal sufficient statistics and the LLM generators and parameters:

Structural Relationship	Loglinear Model	Generating Class	Generators	Minimal Sufficient Statistics
Independence	$\log(\mu_{ij}) = \lambda + \lambda_i^X + \lambda_j^Y$	$[X][Y]$	$[X]$ and $[Y]$	$\{n_{i+}\}$, $\{n_{+j}\}$
Dependence	$\log(\mu_{ij}) = \lambda + \lambda_i^X + \lambda_j^Y + \lambda_{ij}^{XY}$	$[XY]$	$[XY]$	$\{n_{ij}\}$

Table 3.1 summarizes the *df* and maximum likelihood estimates (MLEs) corresponding to the LLMs for the two-way table.

Table 3.1. Summary of the Loglinear Models for the Two-Way Contingency Table

Association Structure	Generating Class	df	Maximum Likelihood Estimate of μ_{ij}, $\hat{\mu}_{ij}$
$[X \otimes Y]$ First-order interaction	$[X][Y]$	$(I-1)(J-1)$	$n_{i+}n_{+j}/n$
between X and Y	$[XY]$	0	n_{ij}

3.2 Three-Way Contingency Table

As was seen in Chapter 2, there are five different kinds of structural association models among three categorical variables. In the following, the LLM is developed for each kind of structural association, paralleling the development in Subsections 2.3.1 to 2.3.5. For simplicity, it is assumed that for each LLM, the λ-terms sum to 0 over the appropriate indices (zero-sum constraints), even though they are not written each time; the ranges of the indices are $i = 1, 2, \ldots, I; j = 1, 2, \ldots, J;$ and $k = 1, 2, \ldots, K$.

3.2.1. Mutual Independence

When the categorical variables X, Y, and Z are mutually independent, the generating class is $[X][Y][Z]$ and the LLM is

$$\log(\mu_{ijk}) = \lambda + \lambda_i^X + \lambda_j^Y + \lambda_k^Z.$$

The factorization of the joint probability for this model,

$$\pi_{ijk} = \pi_{i++}\pi_{+j+}\pi_{++k},$$

can be derived directly from the LLM since $\mu_{ijk} = n\pi_{ijk}$, and so

$$\pi_{ijk} = \frac{1}{n}\exp(\lambda + \lambda_i^X + \lambda_j^Y + \lambda_k^Z).$$

Now take appropriate sums—for example, $\pi_{i++} = (1/n)\exp(\lambda + \lambda_i^X)$ $\sum_{j,k}\exp(\lambda_j^Y + \lambda_k^Z)$, and so on—and use $n = \mu_{+++} = \sum_{i,j,k}\exp(\lambda + \lambda_i^X + \lambda_j^Y + \lambda_k^Z)$ to prove that $\pi_{ijk} = \pi_{i++}\pi_{+j+}\pi_{++k}$. See Bishop et al. (1975, sec. 2.4) for details.

The λ-terms in this model are called the *main effects*:

$$\lambda_i^X = \frac{1}{JK}\sum_{j,k}\log(\mu_{ijk}) - \frac{1}{IJK}\sum_{i,j,k}\log(\mu_{ijk}), \qquad i = 1, 2, \ldots, I,$$

and similarly for λ_j^Y and λ_k^Z.

3.2.2. Joint Independence

Suppose X is *jointly independent* of Y and Z: $[X \otimes Y, Z]$. In this case, the generating class is $[X][YZ]$, and the LLM is

$$\log(\mu_{ijk}) = \lambda + \lambda_i^X + \lambda_j^Y + \lambda_k^Z + \lambda_{jk}^{YZ}.$$

The λ_{jk}^{YZ} term, called the *first-order interaction* between Y and Z, has been added to the mutual independence model to represent the association between Y and Z. The probabilistic model for this case,

$$\pi_{ijk} = \pi_{i++}\pi_{+jk}, \qquad i = 1, 2, \ldots, I; j = 1, 2, \ldots, J;$$
$$\text{and } k = 1, 2, \ldots, K,$$

can be derived directly from the LLM. Analogous LLMs can be written for the cases $[Y \otimes X, Z]$ and $[Z \otimes X, Y]$.

3.2.3. Conditional Independence

Suppose that conditional on the level of Z, X and Y are independent: $[X \otimes Y|Z]$. The generating class for this case is $[XZ,YZ]$, and the LLM is

$$\log(\mu_{ijk}) = \lambda + \lambda_i^X + \lambda_j^Y + \lambda_k^Z + \lambda_{ik}^{XZ} + \lambda_{jk}^{YZ}.$$

The probabilistic model,

$$\pi_{ijk} = \frac{\pi_{i+k}\pi_{+jk}}{\pi_{++k}},$$

can be derived directly from the LLM. In this model, there are two first-order interaction terms: λ_{ik}^{XZ} measures the association between X and Z, while λ_{jk}^{YZ} measures the association between Y and Z. Analogous models can be written for the cases $[X \otimes Z | Y]$ and $[Y \otimes Z | X]$.

3.2.4. Homogeneous Association

The generating class for the homogeneous association model is $[XY]$ $[XZ][YZ]$, and the LLM is

$$\log(\mu_{ijk}) = \lambda + \lambda_i^X + \lambda_j^Y + \lambda_k^Z + \lambda_{ij}^{XY} + \lambda_{ik}^{XZ} + \lambda_{jk}^{YZ}.$$

This model contains all the main effects and pairwise first-order interactions among the variables X, Y, and Z; there are no independencies or conditional independencies among the variables in this model. There is no closed-form factorization of π_{ijk} for this LLM (see Subsection 2.3.4).

3.2.5. Saturated Model

The generating class for this model is $[XYZ]$, and the LLM is

$$\log(\mu_{ijk}) = \lambda + \lambda_i^X + \lambda_j^Y + \lambda_k^Z + \lambda_{ij}^{XY} + \lambda_{ik}^{XZ} + \lambda_{jk}^{YZ} + \lambda_{ijk}^{XYZ}.$$

This model contains all the main effects, all the first-order interactions, and the second-order interaction. The second-order interaction measures the change in the association between any two of the variables with respect to the levels of the third variable. Note that when the association between any two variables does not change across the levels of the third variable, then $\lambda_{ijk}^{XYZ} = 0$ for all i, j, and k and the homogeneous association model obtains. For this reason, the homogeneous association model is sometimes called the "no second-order interaction" model, and the saturated model is sometimes called the "second-order interaction" model.

Table 3.2 provides a summary of the LLMs for the three-way contingency table.

Table 3.2. Summary of the Loglinear Models for the Three-Way Contingency Table

Association Structure	Generating Class	df	Maximum Likelihood Estimate of μ_{ijk}, $\hat{\mu}_{ijk}$
$[X \otimes Y \otimes Z]$	$[X][Y][Z]$	$IJK - I - J - K + 2$	$n_{i++}n_{+j+}n_{++k}/n^2$
$[X \otimes Y, Z]$	$[X][YZ]$	$(I-1)(JK-1)$	$n_{i++}n_{+jk}/n$
$[X \otimes Y Z]$	$[XZ][YZ]$	$K(I-1)(J-1)$	$n_{i+k}n_{+jk}/n_{++k}$
Homogeneous association	$[XY][XZ][YZ]$	$(I-1)(J-1)(K-1)$	[a]
Saturated model	$[XYZ]$	0	n_{ijk}

a. No closed-form factorization for $\hat{\mu}_{ijk}$ in terms of marginal frequencies.

3.3 Relationships Among the LLMs for a Three-Way Table

The five types of LLMs enumerated in Section 3.2 have interrelationships that are described below.

- If X, Y, and Z are mutually independent, then X is jointly independent of Y and Z: $[X \otimes Y \otimes Z] \rightarrow [X \otimes Y, Z]$. (The symbol "$\rightarrow$" denotes "implies.") Similarly, we have $[X \otimes Y \otimes Z] \rightarrow [Y \otimes X, Z]$ and $[X \otimes Y \otimes Z] \rightarrow [Z \otimes X, Y]$. That is, mutual independence implies joint independence.
- If X is jointly independent of Y and Z, then X and Y are independent conditional on Z: $[X \otimes Y, Z] \rightarrow [X \otimes Y|Z]$. Similarly, $[X \otimes Y, Z] \rightarrow [X \otimes Z|Y]$. That is, joint independence implies conditional independence.
- If X is jointly independent of Y and Z, then X and Y are independent in the marginal table: $[X \otimes Y, Z] \rightarrow [X \otimes Y]$. Similarly, $[X \otimes Y, Z] \rightarrow [X \otimes Z]$. That is, joint independence implies marginal independence.
- If X and Y are independent conditional on Z, then X and Y are *not* necessarily independent in the marginal table: $[X \otimes Y|Z]$ does not imply $[X \otimes Y]$.
- If X and Y are independent in the marginal table, then X and Y are *not* necessarily independent conditional on Z: $[X \otimes Y]$ does not imply $[X \otimes Y|Z]$.

To summarize, mutual independence implies joint independence, and joint independence implies both conditional independence and marginal independence, but conditional independence neither implies nor is implied by marginal independence. Briefly put, we have

Mutual independence ⟶ Joint independence ⟨ Conditional independence / Marginal independence

The lack of correspondence between conditional and marginal independence has been the source of much confusion among nonstatisticians. The X-Y conditional association may differ from the X-Y marginal association. In fact, it's possible for the direction of the association to reverse—a condition known as Simpson's paradox (see, e.g., Agresti, 2002, chap. 2). Only under special conditions, called *collapsibility conditions*, is the X-Y conditional association the same as the X-Y marginal association. This is discussed more fully in Chapter 5.

Example 3.3.1. A Gallup Poll of 825 Catholics was taken in 1999 (www.thearda.com/Archive/Files/Codebooks/GALLUP99_CB.asp); the data used in this example involve the following variables:

S = Support of Women in the Priesthood (Yes, No)
B = Use of Birth Control (Yes, No)
G = Gender (M, F)

Table 3.3 shows the three-way contingency table. A backward elimination selection procedure identifies the best-fitting model as $[SB][G]$. Gender is independent of both Support of Women in the Priesthood and Use of Birth Control, but the latter two factors are related. The relevant P values are given in Table 3.4.

Table 3.3. Cross-Classification of 825 Catholics, 1999 Gallup Poll

		B = Use of Birth Control	
G = *Gender*	S = *Support of Women in the Priesthood*	Yes	No
Male	Yes	193	45
	No	67	53
Female	Yes	240	58
	No	110	59

NOTE: See Example 3.3.1.

Table 3.4. Results of Statistical Tests of First- and Higher-Order Interactions for the Data of Example 3.3.1

Effect	H_0	*df*	*P Value*
First-order interaction: $S \times B$	$\lambda_{ij}^{SB} = 0$ for all i and j	1	< .0001
First-order interaction: $S \times G$	$\lambda_{ik}^{SG} = 0$ for all i and k	1	.6904
First-order interaction: $B \times G$	$\lambda_{jk}^{BG} = 0$ for all j and k	1	.2850
Second-order interaction: $S \times B \times G$	$\lambda_{ijk}^{SBG} = 0$ for all i, j, and k	1	.1974

NOTE: S = Support of Women in the Priesthood; B = Use of Birth Control; G = Gender.

For interpretation purposes, consider the following sample odds ratios calculated from the observed cell frequencies. The P values come from Table 3.4. The third variable in each case has been collapsed over. (It is appropriate to do so for this model—see Chapter 5 and Table 5.5.)

Relationship	Sample Odds Ratio	P Value[a]	Interpretation
S-G relationship; collapsed over the levels of B	$\dfrac{\text{Odds}_{\text{male}}(\text{Support})}{\text{Odds}_{\text{female}}(\text{Support})} = \dfrac{1.9833}{1.7633} = 1.12$	.6904	The odds of support of women in the priesthood are comparable for men and women
B-G relationship; collapsed over the levels of S	$\dfrac{\text{Odds}_{\text{male}}(\text{Use birth control})}{\text{Odds}_{\text{female}}(\text{Use birth control})} = \dfrac{2.6531}{2.9915} = 0.89$	.2850	The odds of using birth control are comparable for men and women
S-B relationship; collapsed over the levels of G	$\dfrac{\text{Odds}_{\text{use birth control}}(\text{Support})}{\text{Odds}_{\text{do not use birth control}}(\text{Support})} = \dfrac{2.4463}{0.9196} = 2.66$	< .0001	The odds of support of women in the priesthood is estimated to be 2.66 times higher among those who use birth control than among those who do not use birth control

a. The hypotheses being tested are H_0: Odds ratio $= 1$ versus H_A: Odds ratio $\neq 1$.

Summary: Neither Support of Women in the Priesthood nor Use of Birth Control is related to Gender, but they are related to each other. Those who use birth control are more supportive of women in the priesthood than those who don't use birth control; in fact, the odds of support of women in the priesthood by those who use birth control is 2.7 times higher than among those who do not use birth control.

3.4 LLM and Contingency Table Properties

Various properties of LLMs and of contingency tables are discussed below.

3.4.1. Hierarchical LLMs

Consider a given λ-term, say λ^{XY} (the subscripts are omitted for convenience). Then, the λ-terms λ^X and λ^Y are referred to as *lower-order relatives* of λ^{XY}. In general, for any λ-term λ^Θ, its lower-order relatives are all λ-terms λ^θ, where θ is a proper subset of Θ. Conversely, λ^{XYZ} is a *higher-order relative* of λ^{XY}; in general, λ^Θ is a higher-order relative of λ^θ if θ is a proper subset of Θ.

A *hierarchical* LLM is one in which for any λ-term in the model, all its lower-order relatives are also included in the model: If λ^Θ is contained in the LLM, then λ^θ is contained in the LLM for all $\theta \subset \Theta$. This is referred to as the *hierarchy principle*. For example, the homogeneous association model,

$$\log(\mu_{ijk}) = \lambda + \lambda_i^X + \lambda_j^Y + \lambda_k^Z + \lambda_{ij}^{XY} + \lambda_{ik}^{XZ} + \lambda_{jk}^{YZ},$$

is hierarchical, but the following model is not:

$$\log(\mu_{ijk}) = \lambda + \lambda_i^X + \lambda_j^Y + \lambda_{ij}^{XY} + \lambda_{ik}^{XZ} + \lambda_{jk}^{YZ}.$$

(Note, e.g., that λ^{XZ} is in this model but its lower-order relative λ^Z is not.)

Note that the generating class notation is appropriate only for hierarchical LLMs. That is, if $[XYZ]$ is one of the generators of a given LLM, then the λ^{XYZ} term is in the LLM and, by implication, all lower-order relatives of λ^{XYZ} are in the LLM as well. The hierarchical LLMs are the LLMs that are used most commonly in practical applications. The reason for this is that for nonhierarchical LLMs, the statistical significance and interpretation of a high-order λ-term depends on how the variables are coded; for hierarchical

LLMs, the results do not depend on the coding. As Agresti (2002, p. 317) states, using nonhierarchical LLMs is analogous to using ANOVA (analysis of variance) or regression models with interaction terms but without the corresponding main effects.

3.4.2. Partial Association Model

A particularly important kind of LLM is the *partial association model* (Birch, 1965). This is an LLM in which there is at least one first-order interaction equal to 0. For example, the homogeneous association model, [*XY*] [*XZ*][*YZ*], is not a partial association model because all the first-order interaction terms appear in the model. But the conditional independence model, [*XY*][*XZ*], is a partial association model because the λ^{YZ} terms do not appear in the model. For LLMs that are not partial association models, there are no independencies or conditional independencies among the variables. Thus, it is only for the partial association models that independence or conditional independence between variables will occur.

Note that for hierarchical LLMs if λ^{θ} is set to 0, then all its higher-order relatives must also be set to 0. So a partial association model is obtained if. at least one first-order interaction λ-term is set to 0.

3.4.3. Sampling Designs

There are many ways in which the data for a contingency table can be obtained, including the use of quite complex sampling designs. In a two-way table, the three most basic sampling plans used in practical applications are given below:

- *Poisson sampling plan:* Randomly sample subjects from the study population for a fixed period of time, and then cross-classify them according to the two variables *X* and *Y*. For example, randomly sample pedestrians arriving at a given intersection for 1 hour, and cross-classify them according to ethnic group and gender. In this case, the total sample size, n, is a random variable.

- *Multinomial sampling plan:* Randomly sample n subjects from the study population, where n is a fixed, known positive integer, and then cross-classify them according to the two variables *X* and *Y*. For example, randomly sample 100 Ohioans, and cross-classify them according to ethnic group and gender. In this case, n is a fixed constant determined by the experimenter.

- *Product-multinomial sampling plan:* Randomly sample n_{i+} subjects from level i of variable *X*, with $i = 1, 2, \ldots, I$, where n_{i+} is a fixed, known positive integer for each i, and then classify them according to the levels of variable Y. This is also known as *stratified sampling* by rows. Analogously, one can carry out stratified sampling by columns, classifying the n_{+j} subjects from column j, with $j = 1, 2, \ldots, J$, according to the levels of variable X. For example, randomly sample 100 female Ohioans and 100 male Ohioans, and classify them according to ethnic group.

The inferential procedures for testing the fit of an LLM to a set of contingency table data using the chi-squared or the likelihood ratio statistics are valid for these sampling plans, assuming that the sample size is adequately large. They are not valid, however, for contingency table data that have been obtained by other means. For an example of a contingency table that was obtained by sampling methods other than one of the above sampling plans, see Fienberg (1981, Table 2–4). For a real-data example of such a table and the corresponding incorrect chi-squared analysis, see Stewart, Paris, Pelton, and Garretson (1984).

When working with LLMs, care must be taken to ensure that the model is consistent with the sampling plan used to create the contingency table. When a product-multinomial sampling plan is used, the λ-terms in the LLM whose indices correspond to the margins that are fixed by the sampling plan must be retained in the model so that the observed margins match the MLEs of the fixed margins. For example, if the row margins are fixed in a two-way table by the sampling design, then the λ^X-term must appear in any best-fitting LLM considered. Or suppose that 50 African Americans of each sex and 50 Caucasians of each sex (200 subjects in all) are randomly selected and each is asked if he or she believes in the death penalty. With factors $X =$ Ethnic Group, $Y =$ Gender, and $Z =$ Response (Yes/No), the first-order interaction term λ^{XY} must appear in any best-fitting LLM considered for these data regardless of its statistical significance, since the X-Y margins, $\{n_{ij+}\}$, are fixed by the sampling design: $\{n_{ij+}\} = 50$ for $i = 1, 2$ and $j = 1, 2$. For further discussion, see Bishop et al. (1975, chap. 13).

3.4.4. Complete Contingency Tables

A *complete* three-way contingency table is one in which $\mu_{ijk} > 0$ for all i, j, and k. An *incomplete* table is one in which $\mu_{ijk} = 0$ for at least one cell. The zero frequency that occurs in cells for which $\mu_{ijk} = 0$ is called a *structural zero*. A *sampling zero* occurs when $\mu_{ijk} > 0$ but $n_{ijk} = 0$. So a structural zero corresponds to a set of conditions that are impossible, while a sampling zero corresponds to a set of conditions that are possible but rare. Consider a two-way contingency table for the variables (1) Type of Medical Operation and (2) Gender. The cell (conditions) corresponding to (1) Type of

Medical Operation = Hysterectomy and (2) Gender = Male would contain a structural zero. Consider the two-way table for Age Group by Highest Degree Earned. If a zero appears in the cell corresponding to (1) Age Group = 15 to 24 and (2) Highest Degree Earned = PhD, then it would be a sampling zero because earning a PhD by age 24 is rare but not impossible.

Incomplete contingency tables require a special analysis technique to take into account the structural zeros; special formulas are used to calculate degrees of freedom, *quasi-loglinear* models are used to model the data, and the notion of *quasi-independence* is used to interpret the data. See, for example, Fienberg (1981, chap. 8) and Bishop et al. (1975, chap. 5).

3.5 Multi-Way Tables

The LLM can be extended to four- and higher-dimensional tables. The notions of sampling designs and completeness, the hierarchy principle, and the model selection procedures extend to multi-way tables, though with a higher level of complexity.

In the case of high-dimensional multi-way tables, the LLM becomes more complex. For example, for a 2^K contingency table, the saturated LLM has 2^K terms. As mentioned in Chapter 1, statistical methodologies and software are available to find the best-fitting model for such tables. However, the strategies for interpreting and analyzing the properties of the resulting best-fitting model are less well developed. That is, an organized, methodical, comprehensive, and direct procedure for identifying all the important properties of a given LLM and all the independencies and conditional independencies implied by the LLM is lacking. The graphical procedures in this book will fill that gap.

In the past three decades, it has been discovered that mathematical graph theory can be used to great advantage in accomplishing the goals of interpreting and analyzing LLMs for high-dimensional contingency tables. This is the topic of Chapters 4 through 8.

In the analysis of multi-way contingency tables in the social sciences, there are many modeling frameworks and data structures that are encountered: recursive models, latent-class modeling, association models (log-bilinear), correspondence analysis, repeated measurement and other forms of clustered/correlated categorical data, random- and mixed-effects models, logistic and logit models, matched-pairs data, semisymmetric intraclass contingency table data (Khamis, 1983), variables with ordered categories (see Agresti, 1984), and so on. See Goodman (2007) for a review of many of these topics. In this book, we consider only the interpretation and analysis of a given best-fitting hierarchical LLM containing all the main effects for a complete contingency table generated by a standard sampling plan with the goals of

1. identifying the important properties of the given best-fitting LLM and

2. accurately and reliably determining all the independence and conditional independence relationships among the factors from the given LLM.

The remaining chapters address these goals using mathematical graph theory techniques.

CHAPTER 4. ASSOCIATION GRAPHS
FOR LOGLINEAR MODELS

Mathematical graph theory has been shown to be effective in interpreting and analyzing loglinear models (LLMs). By "interpreting" an LLM, we mean determining which variables are independent or conditionally independent of each other. That is, we wish to identify all independencies and conditional independencies (if any) among the categorical variables. By "analyzing" an LLM, we mean

- determining if the model is decomposable;
- if it is decomposable, obtaining the factorization of the joint probability; and
- identifying the collapsibility conditions.

Why are these three items important? Because

1. decomposable models have special properties that are important from both theoretical and practical points of view (see Section 4.4 for more details);
2. explicit, closed-form expressions of the joint probability are useful in theoretical derivations for statistical methodological research (again, see Section 4.4); and
3. it is crucial in applied work to be aware that collapsing over the levels of a factor may change the relationships among the remaining factors, thereby leading to spurious conclusions.

In the following, the definition of a mathematical graph is given, and then basic graph theory principles are presented and used to define and construct the association graph for the LLMs of two-, three-, and multi-way tables. Finally, the association graph is used to interpret and analyze the LLM.

4.1 Basic Graph Theory Principles

A *mathematical graph* is defined as a set of *vertices* and a set of *edges*. Any two vertices may or may not be joined by an edge.

4.1.1. Construction of the Association Graph

For our purposes, the vertices will be the categorical variables under study, and the edges will be first-order interactions. So an edge joins two vertices X and Y if there is a first-order interaction between the two factors X and Y in the LLM; that is, an edge joins X and Y if X and Y are associated. If X and Y are statistically independent, then they are not joined by an edge in the graph. Such a graph is called an *association graph*. (This name was used by Agresti, 2002; it was called a *first-order interaction graph* by Darroch et al., 1980; Andersen, 1997, called it an *association diagram*.)

Consider the two-way contingency table with variables X and Y. If X and Y are independent (i.e., the generating class for the LLM is $[X][Y]$), then the association graph is

$$X \qquad Y$$

If X and Y are associated (i.e., the generating class for the LLM is $[XY]$), then the association graph is

$$X \text{———} Y$$

4.1.2. Adjacency and Conditional Independence

Two vertices in a graph are said to be *adjacent* if an edge joins them. A *path* from vertex X to vertex Y in a graph is a sequence of edges leading from X to Y. A set S of vertices *separates* the two sets T and V of vertices if all paths from any vertex in T to any vertex in V pass through at least one vertex in S. In the two association graphs below, let S = $\{C\}$, T = $\{A, B\}$, and V = $\{D, E\}$. Then, for the graph on the left, S separates T and V, but for the graph on the right, S does not separate T and V because there is a path from a vertex in V (viz., E) to a vertex in T (viz., B) that does not go through S.

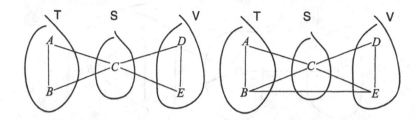

The relationship between adjacency in the association graph and conditional independence between variables is developed in Darroch et al. (1980), using the so-called Markov random fields. The authors show that two variables (vertices in the association graph) are nonadjacent if and only if they are *conditionally independent* relative to all the other variables. If two variables are adjacent, they are said to be *partially associated* relative to the other variables. If a set S of variables separates two sets T and V of variables in the association graph, then the variables in T are conditionally independent of the variables in V relative to the variables in S. As an example, in the association graph on the left above, A and B are independent of D and E conditional on C. For the association graph on the right, if we select S = $\{B, C, D\}$, T = $\{A\}$, and V = $\{E\}$ (see below), then A is independent of E conditional on B, C, and D.

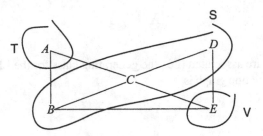

Note that these choices of S, T, and V, where S separates T and V, are not at all unique. Other choices could be made, thereby leading to other conditional independencies. For instance, in the association graph above, we could let S = $\{A, C, E\}$, leading to the independence between B and D conditional on A, C, and E.

In the association graph below, the set of variables V = $\{A, B, C\}$ is independent of T = $\{F\}$ conditional on S = $\{D\}$ because D separates F from A, B, and C. Note also that F is independent of B conditional on A, C, and D (S = $\{A, C, D\}$).

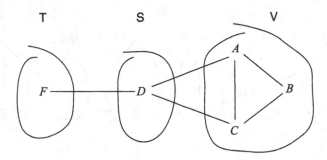

Unconditionally independent variables are said to lie in different connected components of the association graph. For example, in the association graph below, variables X, Y, and Z are independent of variables V and W:

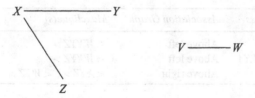

4.1.3. Graphical Models

A *complete graph* has an edge joining every pair of vertices. A *maxclique* in a graph is a maximal set of variables contained in a complete subgraph. Or a maxclique is a set of vertices that form a complete graph that is not contained in a larger complete graph. For example, the graph on the left below is a complete graph because all possible pairs of vertices have an edge. Thus, this graph has one maxclique: $< WXYZ >$. The graph on the right below is not complete (the edge joining X and W is missing) but has two maxcliques: $< XYZ >$ and $< WYZ >$.

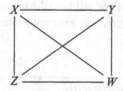 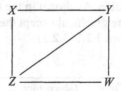

A hierarchical LLM is called a *graphical model* (introduced by Darroch et al., 1980) when there is a one-to-one correspondence between the maxcliques in the association graph and the generators of the LLM. More formally, we say that a model is graphical if whenever the model contains all the first-order interaction terms generated by a higher-order interaction term, then the model also contains the higher-order interaction term. As an example, consider the generating class [$WXYZ$]. The LLM having this generating class is a graphical model because its association graph (above left) has one maxclique, $< WXYZ >$, corresponding directly to the generator of the model, [$WXYZ$]. The LLM having generating class [XYZ][WYZ][WX] has the same association graph (above left) but is not graphical because the maxclique $< WXYZ >$ does not correspond to the generators of the model. The LLM having generating class [XYZ][WYZ] is graphical because its association graph (above right) has maxcliques $< XYZ >$ and $< WYZ >$

corresponding directly to the generators $[XYZ]$ and $[WYZ]$, respectively. A summary of these examples is given below:

Generating Class	Association Graph	Maxclique(s)	Graphical?
$[WXYZ]$	Above left	$<WXYZ>$	Yes
$[XYZ][WYZ][WX]$	Above left	$<WXYZ>$	No
$[XYZ][WYZ]$	Above right	$<XYZ>$, $<WYZ>$	Yes

Graphical models are important because the maxcliques are directly related to the minimal sufficient statistics of the LLM parameters, which can lead to valuable data reduction (see Edwards & Kreiner, 1983, p. 563). In fact, there are statistical techniques for selecting the appropriate graphical models for a given observed contingency table (see Edwards & Kreiner, 1983, sec. 5). With respect to interpretation of the data, the structural associations among the variables in a graphical model (including all second- and higher-order interactions) are immediately apparent from the graph; in particular, nonsaturated graphical models can be interpreted exclusively in terms of independence or conditional independence, and this can be read directly from the graph. Consider, as an example, the three-way table. As shown in the table below, all the LLMs for the three-way table are graphical except the homogeneous association model (refer to Subsections 3.2.1–3.2.5).

Generating Class	Generators	Maxclique(s)	Graphical?
$[X][Y][Z]$	$[X], [Y], [Z]$	$<X>$, $<Y>$, $<Z>$	Yes
$[X][YZ]$	$[X], [YZ]$	$<X>$, $<YZ>$	Yes
$[XZ][YZ]$	$[XZ], [YZ]$	$<XZ>$, $<YZ>$	Yes
$[XY][XZ][YZ]$	$[XY], [XZ], [YZ]$	$<XYZ>$	No
$[XYZ]$	$[XYZ]$	$<XYZ>$	Yes

All nonsaturated graphical models are interpreted directly in terms of independence or conditional independence (see Subsections 3.2.1–3.2.4). The homogeneous association model (nongraphical), however, cannot be thus interpreted; rather, it must be interpreted in terms of the absence of a second-order interaction effect. For example, one may write, "The association between any two factors is homogeneous across the levels of a third factor." The saturated model, which is always graphical, is a special case in which all the interactions are nonzero.

The second reason why graphical models have practical importance is that they play an important role in the identification of the decomposability of an LLM (see Subsection 4.1.5 below).

4.1.4. Chordal Graphs

A *cycle* is a sequence of edges that begins and ends at a given vertex. A *chord* is an edge between nonconsecutive vertices along a cycle. A *chordal graph* is a graph in which every cycle of length four or more has a chord (see Blair & Peyton, 1993; Golumbic, 1980, chap. 4). Linear-time "maximum-cardinality search" algorithms are available to test whether a graph is chordal (Tarjan & Yannakakis, 1984). The graph on the left below is not chordal, because it has a cycle of length four without a chord (A-B-C-D). The graph on the right is chordal.

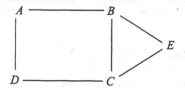 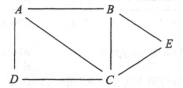

Note that, trivially, any graph with three or fewer vertices is chordal since there is no cycle of length four or more and hence no cycle of length four or more without a chord.

4.1.5. Decomposable LLMs

Darroch et al. (1980) characterize decomposable models as those graphical models whose association graph is a chordal graph; briefly put, "graphical + chordal = decomposable." Consider the LLM having generating class [AB][BC][CD][AD][BCE]. The association graph is on the left below. This LLM is graphical because the generators, [AB], [BC], [CD], [AD], and [BCE], correspond directly to the maxcliques in the association graph, < AB >, < BC >, < CD >, < AD >, and < BCE >, respectively. But the association graph is not chordal, because there's a cycle of length four (A-B-C-D) without a chord. Therefore, this model is not decomposable.

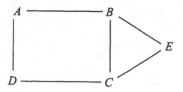 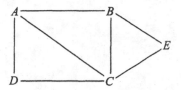

Consider the LLM having generating class [ABC][ACD][BCE]. The association graph is on the right above. This model is graphical because the generators, [ABC], [ACD], and [BCE], correspond directly to the max-cliques of the association graph, $<ABC>$, $<ACD>$, and $<BCE>$, respectively. The association graph is chordal because there is no cycle of length four or more without a chord. Thus, this LLM is decomposable. Consider the LLM having generating class [AB][BC][AC][AD][CD][BCE]. The association graph is on the right above. The association graph is chordal, but the model is not graphical, because the maxclique $<ABC>$ does not have a corresponding generator, [ABC], in the generating class. A summary of these three examples is given below:

Generating Class	Association Graph	Graphical LLM?	Chordal Association Graph?	Decomposable?
[AB][BC][CD][AD][BCE]	Above left	Yes	No	No
[ABC][ACD][BCE]	Above right	Yes	Yes	Yes
[AB][BC][AC][AD][CD][BCE]	Above right	No	Yes	No

Now these definitions will be applied to LLMs for three-way contingency tables to solidify the concepts.

4.2 Association Graphs for Three-Way Tables

In the following, the association graph will be given for each of the five different kinds of structural association models for three categorical variables presented in Section 3.2. In each case, an edge joins two vertices (variables) if there is a first-order interaction between the two variables—that is, if there is a first-order interaction λ-term (double-subscripted λ-term) in the LLM.

4.2.1. Mutual Independence: $log(\mu_{ijk}) = \lambda + \lambda_i^X + \lambda_j^Y + \lambda_k^Z$

The generating class is [X][Y][Z]. The association graph is

X

Y Z

There are no first-order interactions in the mutual independence model, so there are no edges in the association graph. In this case, each variable is a separate (nonadjacent) component of the graph, indicating unconditional independence among all three variables. The maxcliques in this graph are $<X>$, $<Y>$, and $<Z>$. These correspond directly to the model generators, $[X]$, $[Y]$, and $[Z]$, respectively, so this is a graphical model.

4.2.2. Joint Independence: $log(\mu_{ijk}) = \lambda + \lambda_i^X + \lambda_j^Y + \lambda_k^Z + \lambda_{jk}^{YZ}$

Here, X is *jointly independent* of Y and Z: $[X \otimes Y, Z]$. The generating class is $[X][YZ]$. The association graph is

$$X$$

$$Y \text{———} Z$$

There is only one first-order interaction term in the LLM, namely, λ_{jk}^{YZ}, hence there is only one edge, joining Y and Z, in the association graph. This association graph consists of two components: (1) X is in one and (2) Y and Z are in the other. So X is unconditionally independent of Y and Z. The maxcliques in this graph are $<X>$ and $<YZ>$. These correspond directly to the generators, $[X]$ and $[YZ]$, respectively, therefore this is a graphical model.

4.2.3. Conditional Independence:
$log(\mu_{ijk}) = \lambda + \lambda_i^X + \lambda_j^Y + \lambda_k^Z + \lambda_{ik}^{XZ} + \lambda_{jk}^{YZ}$

Conditional on the level of Z, X and Y are independent: $[X \otimes Y|Z]$. The generating class is $[XZ][YZ]$. The association graph is

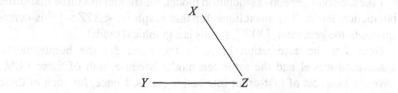

$$X$$

$$Y \text{———} Z$$

There are two first-order interaction terms in this LLM: λ_{ik}^{XZ} and λ_{jk}^{YZ}; hence, there are two edges, one joining X and Z and the other joining Y and Z. In this association graph, the variable Z separates X and Y. So X is independent of Y conditional on Z. The maxcliques in this graph are $<XZ>$ and $<YZ>$. These correspond to the generators, $[XZ]$ and $[YZ]$, respectively, so this is a graphical model.

4.2.4. Homogeneous Association:

$$log(\mu_{ijk}) = \lambda + \lambda_i^X + \lambda_j^Y + \lambda_k^Z + \lambda_{ij}^{XY} + \lambda_{ik}^{XZ} + \lambda_{jk}^{YZ}$$

The generating class is $[XY][XZ][YZ]$. The association graph is

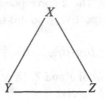

There are three first-order interaction terms in this LLM: λ_{ij}^{XY}, λ_{ik}^{XZ}, and λ_{jk}^{YZ}; hence, there are three edges. The maxclique in this graph is $<XYZ>$; since there is no generator $[XYZ]$ in the generating class, this is not a graphical model.

4.2.5. Saturated Model:

$$log(\mu_{ijk}) = \lambda + \lambda_i^X + \lambda_j^Y + \lambda_k^Z + \lambda_{ij}^{XY} + \lambda_{ik}^{XZ} + \lambda_{jk}^{YZ} + \lambda_{ijk}^{XYZ}$$

The generating class is $[XYZ]$. The association graph is

Like the homogeneous association model, this model has three first-order interaction terms. The maxclique for this graph is $<XYZ>$; this corresponds to the generator, $[XYZ]$, so this is a graphical model.

Note that the association graph is the same for the homogeneous association model and the saturated model because both of these LLMs have the same set of first-order interaction terms. Hence, for each of these LLMs, the association graph has one maxclique, $<XYZ>$. However, the homogeneous association model is not graphical because the maxclique, $<XYZ>$, does not have a one-to-one correspondence with the generators of the model, $[XY]$, $[XZ]$, and $[YZ]$. The saturated model, however, is graphical because the maxclique, $<XYZ>$, corresponds directly to the generator of the model, $[XYZ]$. (See Subsection 4.1.3 for another example of two different generating classes having the same association graph.)

Since the above models involve only three variables, the corresponding association graphs are trivially chordal (see Subsection 4.1.4). Because the homogeneous association model is not graphical, it is not decomposable. All the other models above are decomposable because each of them is graphical and each has an association graph that is chordal. Recall that "graphical + chordal = decomposable." These conclusions are consistent with the results in Table 3.2, where it is shown that the joint distribution can be factored explicitly (a consequence of decomposability) for all the above models except the homogeneous association model.

4.3 Association Graphs for Multi-Way Tables

For four-way tables and higher, the above principles extend straightforwardly. Eight examples will be given below to illustrate this. The same eight examples will be used in succeeding chapters to illustrate a variety of techniques.

Example 4.3.1. Consider the LLM having generating class [*ABC*][*BD*] [*CD*]. The association graph is

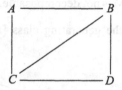

In this graph, there are two maxcliques: $< ABC >$ and $< BCD >$. These do not correspond to the generators of the model ([*BCD*] is not a generator), so this LLM is not graphical. This graph is chordal because the cycle of length four, $< ABCD >$, has a chord, $< BC >$. However, this LLM is not decomposable because decomposability requires that the association graph be chordal *and* the LLM be graphical.

Example 4.3.2. Consider the LLM having generating class [*ABC*][*BD*]. The association graph is

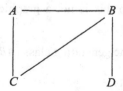

There are two maxcliques in this graph: $<ABC>$ and $<BD>$; these match the generators of the model, so this is a graphical model. There is no cycle of length four in this graph (only one of length three, $<ABC>$), so trivially, this graph is chordal. Since this is a graphical model having a chordal association graph, the LLM is decomposable.

Example 4.3.3. Consider the generating class $[AB][BD][CD][AC]$. The association graph is

There are four maxcliques in this graph: $<AB>$, $<BD>$, $<CD>$, and $<AC>$. These correspond exactly with the model generators, so this LLM is graphical. However, the association graph has a cycle of length four without a chord, so the LLM is not decomposable.

Example 4.3.4. Consider the generating class $[AS][ACR][MCS][MAC]$. The association graph is

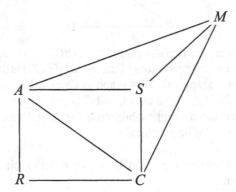

Note that $<MACS>$ is a maxclique but there is no generator $[MACS]$ in the generating class, so this is not a graphical model and hence not decomposable.

Example 4.3.5. Consider the generating class $[ABCD][ACE][BCG][CDF]$. The association graph is

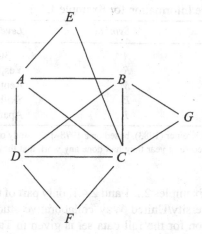

This model is graphical, and the association graph is chordal, so the LLM is decomposable.

Example 4.3.6. Edwards and Kreiner (1983) analyzed a set of data (unpublished) from an investigation conducted at the Institute for Social Research, Copenhagen, collected during 1978 and 1979. A sample of 1,592 employed men, 18 to 67 years old, were asked whether in the preceding year they had done any work that they would formerly have paid a craftsman to do. The variables are given in Table 4.1. The purpose of the investigation was, among other things, to estimate the extent of tax evasion in the building industry. One of the LLMs that the authors considered was [ARME][AMET]. The association graph for this model is

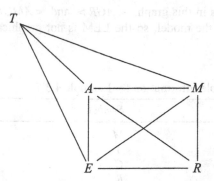

Note that the maxcliques match the generators of the model and that the association graph is chordal, so this LLM is decomposable.

Table 4.1. Variable Information for Example 4.3.6

Variable	Symbol	Levels
Age Category (years)	A	< 30, 31–45, 46–67
Response	R	Yes, No
Mode of Residence	M	Rented, Owned
Employment	E	Skilled, Unskilled, Other
Type of Residence	T	Apartment, House

SOURCE: Edwards and Kreiner (1983), based on a 1978–1979 study of 1,592 employed men asked whether in the preceding year they had done any work that they would formerly have paid a craftsman to do.

Example 4.3.7. In Examples 2.2.1 and 2.4.1, only part of the data set for the Wright State University/United Way collaborative study was used. The variable specification for the full data set is given in Table 4.2. Here, we have A = Alcohol Use, C = Cigarette Use, M = Marijuana Use, G = Gender, and R = Race. The analysis of these data was carried out by the Statistical Consulting Center at Wright State University, Dayton, Ohio (see Agresti, 2002, sec. 9.2.2, for a comprehensive analysis of the model fitting). One model considered by Agresti (2002, p. 363) has generating class $[AC][AM][CM][AG][AR][GR]$. The association graph for this model is

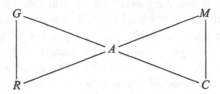

The maxcliques in this graph, $<AGR>$ and $<AMC>$, do not match the generators of the model, so the LLM is not graphical and hence not decomposable.

Table 4.2. Variable Information for Example 4.3.7

Variable	Symbol	Levels
Marijuana Use	M	No, Yes
Alcohol Use	A	No, Yes
Cigarette Use	C	No, Yes
Race	R	Other, White
Gender	G	Female, Male

SOURCE: Based on a 1992 study of 2,276 nonurban seniors (Grade 12) asked if they had ever used marijuana, alcohol, or cigarettes.

Example 4.3.8. Consider the generating class (using the numbers 0 through 9 to represent 10 categorical variables) [67][013][125][178] [1347][1457] [1479]. This generating class was introduced in Chapter 1 as a motivating example. The association graph is

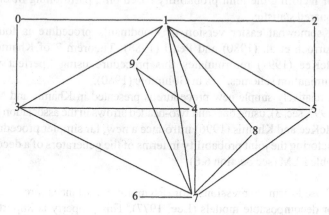

It is not easy to determine from this association graph whether the LLM is decomposable. We will see in Chapter 6, using an alternative graphical procedure, that this model is indeed decomposable (see Example 6.3.8).

4.4 Decomposable LLMs

Why is it important to determine if a given LLM is decomposable? There are several reasons, as given below:

1. As was mentioned in Subsection 2.3.4, the joint probability of the contingency table can be factored in closed form with respect to the indices in the generators of the model only for decomposable models, and for this reason, maximum likelihood estimators are available in closed form only for decomposable models. An iterative fitting algorithm must be used to obtain the maximum likelihood estimates for nondecomposable models. However, given the power and speed of today's computers, this is not a serious practical disadvantage of nondecomposable models. Thus, it may seem that this particular property of decomposable models is irrelevant. However, it becomes quite important in research involving contingency table analysis methodology—see Points 2 and 3 below.

By way of historical development, the method of factoring the joint probability for decomposable LLMs has gone from extremely complex to quite simple, reflecting the growth and development of mathematical/statistical theory and methods over the past 40 years.

- Goodman (1971b) presents a very involved, complicated procedure for factoring the joint probability based on a partitioning of the chi-squared statistic.
- A somewhat easier version of Goodman's procedure is found in Darroch et al. (1980) and Pearl (1988); Theorem 2 of Khamis and McKee (1997) reformulates this procedure using "perfect vertex elimination schemes," as in Golumbic (1980).
- A relatively simple new procedure is presented in Khamis and McKee (1997, sec. 3), using one- and two-headed arrows in the association graph.
- McKee and Khamis (1996) introduce a new, far simpler procedure for factoring the joint probability in terms of the generators of a decomposable LLM (see Section 6.6).

2. Closed-form expressions for asymptotic variances are available only for decomposable models (Lee, 1977). This property is important in theoretical and methodological research, for example, in the study of large, sparse contingency tables (see, e.g., Fienberg, 1979; Koehler, 1986).

3. The conditional G^2 (likelihood ratio) statistic simplifies for decomposable models. This is important because of the extensive use of conditional G^2 statistics in fitting LLMs to data, especially in the case of very sparse contingency tables (see, e.g., Whittaker, 1990), and in testing for marginal homogeneity and interactions.

4. Decomposable LLMs are hierarchical models that are both graphical and *recursive* (a special kind of so-called path analysis model introduced by Goodman, 1973; see also Wermuth, 1980; Wermuth & Lauritzen, 1983). Theorem 4 in Khamis and McKee (1997, sec. 3) shows that decomposable models are graphical models that can be oriented so as to have a causal interpretation. Roughly, the edges of the corresponding chordal graph can be oriented so as to correspond to a causal interpretation. Such models are very important in social science research.

5. Decomposable models are easier to interpret and analyze than nondecomposable models. This will be seen clearly in Chapter 6 (Section 6.7).

Referring to the five points above, it can be seen that some of the advantages of decomposable models are of principal value to researchers in statistical methodology and analysis techniques (see Points 1–3 above). For the social science researcher and other data analysts, the practical advantages of decomposable models are their relationship to recursive models (Point 4 above) and their relative ease of interpretation (Point 5 above).

4.5 Summary

The association graph, or first-order interaction graph, for a hierarchical LLM has vertices = variables and edges = first-order interactions. Four classes of LLMs have been identified:

I. *LLMs* (no restrictions)

II. *Hierarchical LLMs:*

$$\lambda^\Theta \in LLM \text{ implies } \lambda^\theta \in LLM \text{ for all } \theta \subset \Theta$$

III. *Graphical LLMs:*

$$\text{Maxcliques in association graph} \equiv LLM \text{ generators}$$

IV. *Decomposable LLMs:* A graphical model whose association graph is chordal

These classes of LLMs are nested within each other as follows: IV ⊂ III ⊂ II ⊂ I (see Figure 4.1).

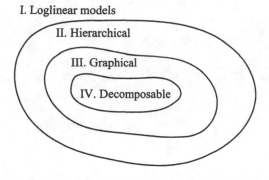

I. Loglinear models

II. Hierarchical

III. Graphical

IV. Decomposable

Figure 4.1. Nested Classes of Loglinear Models

Examples of LLMs that are a member of a given class but not of a smaller class are given below:

I. $\log(\mu_{ijkl}) = \lambda + \lambda_i^X + \lambda_j^Y + \lambda_k^Z + \lambda_l^W + \lambda_{ij}^{XY} + \lambda_{ijk}^{XYZ}$: This model is nonhierarchical.

II. LLM with generating class $[XY][XZ][YZ]$: This LLM is hierarchical, but the generators do not match the maxcliques of the association graph, hence this model is not graphical (see Subsection 4.2.4).

III. LLM with generating class $[AB][BD][CD][AC]$: This is a graphical model, but the association graph is not chordal, hence this model is not decomposable (see Example 4.3.3).

IV. LLM with generating class $[X][Y][Z]$: This model is decomposable.

Note that for the application of the association graph being used here, it makes no difference *how* the graph is drawn. That is, the vertices may be placed in any orientation. However, with a little experience it will be seen that some orientations are easier to work with than others in terms of identifying chordality and maxcliques, especially when there are many vertices and edges with which to contend.

For a general review of the graphical model literature as of 1996, see Khamis and McKee (1997). In the next chapter, collapsibility conditions and how they relate to the association graph will be discussed.

CHAPTER 5. COLLAPSIBILITY CONDITIONS AND THE ASSOCIATION GRAPH

The collapsibility theorem states which relationships are changed and which are unchanged when the levels of one or more variables in a multiway contingency table are collapsed over. In the following, the association graph is seen to facilitate the characterization of the collapsibility theorem in a very easy, straightforward manner.

5.1 Collapsing in a Three-Way Contingency Table

A very common practice among nonstatisticians in the analysis of a three-way contingency table is to collapse (or add) over the levels of one factor, say Z, in order to study the association between the other two factors, say X and Y, in the resulting two-way marginal table. It is tempting to do this because (a) it increases the observed cell frequencies in the X-Y marginal table, thereby producing more stable estimates, and (b) it simplifies the analysis and interpretation. However, as was emphasized in Section 3.3, the association between two variables X and Y as measured by $\left\{ \lambda_{ij}^{XY} \right\}$ may or may not be the same in the X-Y marginal table as it is in the X-Y partial tables. To illustrate this point, consider the following example.

Example 5.1.1. Rodenhauser, Schwenkner, and Khamis (1987) studied the factors related to drug treatment refusal for patients at the Dayton Mental Health Center. To determine if there is a relationship between refusal and a diagnosis of psychosis, the 2×2 table given in Table 5.1 was studied.

Table 5.1. Cross-Classification of 421 Dayton Mental Health Clinic Patients According to Medication Refusal and Diagnosis of Psychosis

	Diagnosed Psychotic?	
Refused Medication?	Yes	No
Yes	115	32
No	192	82

SOURCE: Rodenhauser et al. (1987).

NOTE: See Example 5.1.1.

For this table, the estimated odds ratio is $\hat{\alpha} = 1.53$, with 95% confidence interval [0.96, 2.5]. There is not quite strong enough evidence at the .05 level of significance to claim an association between medication refusal and psychosis.

If these data are broken down by those who have a diagnosed personality disorder and those who don't, one obtains the $2 \times 2 \times 2$ table given in Table 5.2. The sample odds ratios for the two personality disorder groups are $\hat{\alpha}_{personality\,disorder} = 1.04$ (95% confidence interval [0.6, 1.8]) and $\hat{\alpha}_{no\,personality\,disorder} = 5.6$ (95% confidence interval [1.6, 19.5]). The three-way table presents quite a different picture of the refusal-psychosis relationship: For those with a personality disorder there is no strong relationship between medication refusal and a diagnosis of psychosis, but for those without a personality disorder there is a statistically significant relationship (the sample odds of refusing medication are 5.6 times higher for diagnosed psychotics than for those not diagnosed as psychotic).

Table 5.2. Cross-Classification of 421 Dayton Mental Health Clinic Patients According to Diagnosis of Personality Disorder, Medication Refusal, and Diagnosis of Psychosis

Diagnosis of Personality Disorder?	Refused Medication?	Diagnosed Psychotic?	Frequency
Yes	Yes	Yes	54
		No	29
	No	Yes	102
		No	57
No	Yes	Yes	61
		No	3
	No	Yes	90
		No	25

SOURCE: Rodenhauser et al. (1987).

NOTE: See Example 5.1.1.

The above example illustrates that a relationship in the marginal table can be quite different from what it is in the partial tables. Whether the X-Y partial association is preserved in the marginal table depends on the structural associations among X, Y, and Z. In fact, it is possible for the direction of the association in the partial tables to *reverse* in the marginal table. Such a phenomenon is called *Simpson's paradox* (see Simpson, 1951; see Good & Mittal, 1987, for a history).

Even the columnist Marilyn vos Savant, who reportedly has the highest recorded IQ, became confused by a Simpson's paradox problem. In her weekly column, she presented the following artificial tables (the cure rates are in parentheses):

		TRIAL 1		TRIAL 2	
		CURED?		CURED?	
		Yes	No	Yes	No
TREATMENT	A	40 (.20)	160	85 (.85)	15
	B	30 (.15)	170	300 (.75)	100

		Combine TRIALs 1 and 2	
		CURED?	
		Yes	No
TREATMENT	A	125 (.42)	175
	B	330 (.55)	270

Which treatment is better in terms of cure rate? Marilyn vos Savant (1996) responded that Treatment B is better because the cure rate is higher for Treatment B in the combined table. However, the combined table was formed in violation of collapsibility conditions and hence produced spurious results. In fact, Treatment A is better because it has a higher cure rate in both trials. This is an illustration of Simpson's paradox.

A real-data demonstration of this is given in the following example.

Example 5.1.2. Radelet and Pierce (1991) provided the data found in Table 5.3. The data pertain to 674 defendants involved in indictments in murder cases in Florida between 1976 and 1987. (Rather than carrying out a formal, sophisticated inferential procedure, we will merely scrutinize the sample odds ratios in the following discussion to illustrate the point.) For white and black victims separately, the association between the defendant's race and receipt of the death penalty, as measured by the estimated odds ratio, is $\hat{\alpha}_{white\,victim} = 0.43$ and $\hat{\alpha}_{black\,victim} = 0.0$, showing that black defendants receive the death penalty proportionately more often than white defendants regardless of the victim's race. The table obtained by collapsing over victim's race is Table 5.4. In the collapsed table, $\hat{\alpha} = 1.45$, showing that whites receive the death penalty proportionately more often than blacks. Which is the truth? The truth is that black defendants receive the death penalty proportionately more often than whites. To obtain Table 5.4, Table 5.3 was collapsed in violation of collapsibility conditions.

Table 5.3. Cross-Classification of 674 Defendants in Murder Cases According to the Victim's Race, the Defendant's Race, and Whether the Death Sentence Was Given

Victim's Race	Defendant's Race	Received Death Penalty?	Frequency
White	White	Yes	53
		No	414
	Black	Yes	11
		No	37
Black	White	Yes	0
		No	16
	Black	Yes	4
		No	139

Source: Radelet and Pierce (1991).

Note: See Example 5.1.2.

Table 5.4. Marginal Table Obtained by Collapsing Over Victim's Race in Table 5.3

Defendant's Race	Received Death Penalty?	Frequency
White	Yes	53
	No	430
Black	Yes	15
	No	176

Other real-data examples of Simpson's paradox can be found in Wagner (1982) and Agresti (2002, chap. 2).

5.2 The Collapsibility Theorem and the Association Graph

In this chapter, we will be concerned with *parametric collapsibility*: Variables whose categories are summed over are said to be *collapsible* with respect to specific loglinear model (LLM) λ-terms when the specified λ-terms in the original array are identical to the same λ-terms in the corresponding LLM for the reduced array, that is, when the partial associations are equal to the marginal association for the corresponding variables. (This is in contrast to "P collapsibility," which is defined in terms of probabilities and their invariance under collapsing—see Asmussen & Edwards, 1983, for necessary and sufficient conditions for P collapsibility in terms of the association graph; see also Khamis & McKee, 1997, sec. 1.4). As an example of parametric collapsibility, consider the joint independence model $[X][YZ]$, with LLM

$$\log(\mu_{ijk}) = \lambda + \lambda_i^X + \lambda_j^Y + \lambda_k^Z + \lambda_{jk}^{YZ}.$$

As will be seen shortly, the variable X is collapsible with respect to the $\left\{\lambda_{jk}^{YZ}\right\}$ terms. That is, the $\left\{\lambda_{jk}^{YZ}\right\}$ terms in the above model are identical to the $\left\{\lambda_{jk}^{YZ}\right\}$ terms in the LLM for the two-way contingency table obtained by collapsing over the levels of X,

$$\log(\mu_{jk}) = \lambda + \lambda_j^Y + \lambda_k^Z + \lambda_{jk}^{YZ}.$$

The $\left\{\lambda_{jk}^{YZ}\right\}$ terms measure the Y-Z association using odds ratios, so we say that the association between Y and Z in the three-way table (partial association) is the same as in the Y-Z two-way table (marginal association). An example in which the partial association is not the same as the marginal association is given by the homogeneous association model, $[XY][XZ][YZ]$. If we collapse over the levels of X in this model, then the $\left\{\lambda_{jk}^{YZ}\right\}$ terms may change in the reduced table. Hence, X is not collapsible with respect to the Y-Z association. (For further discussion and proofs, see Bishop et al., 1975, Sec. 2.4.)

Collapsibility Theorem

The conditions under which a variable is collapsible are given in a theorem called the *collapsibility theorem*. For each of the five types of structural associations that can exist for three variables (see Subsections 3.2.1–3.2.5), the collapsibility conditions are given in Table 5.5. Note that no first-order interaction λ-terms are invariant to collapsing for the homogeneous association and saturated models. However, all first-order interaction λ-terms are invariant to collapsing for the joint independence and mutual independence models.

Table 5.5. Parametric Collapsibility Conditions for Three Categorical Variables X, Y, and Z

Model	Collapse Over Levels of	Associations Preserved
$[X][Y][Z]$	a	a
$[X][YZ]$	a	a
$[XZ][YZ]$	X	Y-Z
	Y	X-Z
	Z	None
$[XY][XZ][YZ]$	b	b
$[XYZ]$	b	b

a. The levels of any variable can be collapsed over without affecting the association between the other two variables in the resulting marginal table.

b. Collapsing over the levels of any variable may distort the association between the other two variables in the resulting marginal table.

According to Bishop (1971),

> If a table is collapsed by adding over a variable that is correlated with k other variables, then k-factor and lower-order effects relating these variables may be changed in the reduced table. Conversely, effects involving the remaining variables are not affected by collapsing. (p. 549)

With regard to the association graph, this can be translated into the following (see also Agresti, 2002, p. 360):

> Partition the variables (the vertices in the association graph) into three pairwise disjoint subsets S, T, and V such that S separates T and V (see Subsection 4.1.2). Then, on collapsing over the variables in T, all λ-terms relating the variables in V and all λ-terms connecting the variables in S to the variables in V will be unchanged.

Recall from Subsection 4.1.2 that separation in the association graph corresponds to conditional independence. The set of k variables mentioned in Bishop's (1971) statement above is the set of variables called S. This set, S, separates the two sets T and V. Hence, the variables in T are independent of the variables in V conditional on the variables in S. Briefly put,

$$\text{Conditional independence} \rightarrow \text{Collapsibility.}$$

Note that, according to the above conditions, the λ-terms involving the variables exclusively in S are not invariant to collapsing. A few illustrative examples are given below.

Example 5.2.1. Consider the conditional independence model, [XZ][YZ]. The association graph is given below:

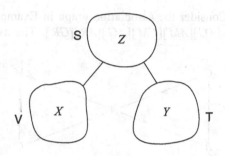

Note that Z separates X and Y: $\mathsf{S} = \{Z\}$, $\mathsf{V} = \{X\}$, and $\mathsf{T} = \{Y\}$. Then one can collapse over the levels of Y without affecting λ^X or λ^{XZ}. One can also collapse over X without affecting λ^Y or λ^{YZ}. That is, because X is independent of Y conditional on Z, the X levels can be collapsed over without affecting the Y-Z parametric association. Similarly, the Y levels can be collapsed over without affecting the X-Z parametric association.

Example 5.2.2. Consider the LLM in Example 4.3.6, having generating class [ARME][AMET]. The association graph is given below along with the partitioning. Here, we can collapse over R and preserve the associations between T and the other variables (λ^{TA}, λ^{TE}, λ^{TM}, λ^{TAM}, etc.). We can also collapse over T and preserve the associations between R and the other variables (λ^{RM}, λ^{RA}, etc.). However, the associations among A, M, and E are not preserved; that is, λ^A, λ^M, λ^E, λ^{AM}, λ^{AE}, λ^{ME}, and λ^{AME} are not invariant to the collapsing.

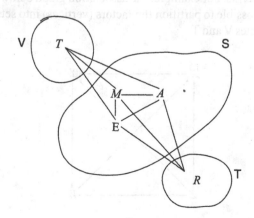

Note that the partitioning of the association graph variables may not be unique. Consider the following example.

Example 5.2.3. Consider the association graph in Example 4.3.7, having generating class $[AC][AM][CM][AG][AR][GR]$. The association graph for this model is

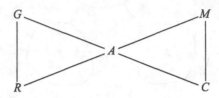

Here, we may choose $S = \{A\}$, $V = \{G, R\}$, and $T = \{M, C\}$. Then one may collapse over M and C, preserving λ^{AG}, λ^{AR}, and λ^{GR}, or one may collapse over G and R, preserving λ^{AC}, λ^{AM}, and λ^{CM}. Alternatively, one may choose $S = \{A, M\}$, $V = \{G, R\}$, and $T = \{C\}$. Then one may collapse over C or over G and R, preserving all other first-order interaction λ-terms *except* λ^{AM}.

5.3 Conclusion

Several consequences of the collapsibility theorem and how they relate to the association graph are described below:

- *Nonpartial association model:* If all the first-order interaction terms are present in the LLM, then the association graph is a complete graph and no separation is possible. Hence, collapsing over any variable may affect all the other λ-terms. For example, the association graph below is complete, and so it is impossible to partition the factors (vertices) into sets S, V, and T so that S separates V and T.

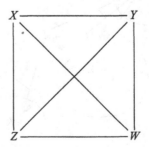

- *Partial association model*: If the first-order interaction term for two variables is missing in the LLM (which means that all the higher-order relatives are missing by the hierarchy principle), then the edge joining those two variables is missing in the association graph. In this case, the two variables can be separated by the other variables in the model. As an example, in the graph above the edge joining X and W is removed, producing the graph below. This graph shows that X and W are independent conditional on Y and Z. Thus, X may be collapsed over without affecting λ^{WZ}, λ^{WY}, or λ^{WYZ}; or W may be collapsed over without affecting λ^{XZ}, λ^{XY}, or λ^{XYZ}.

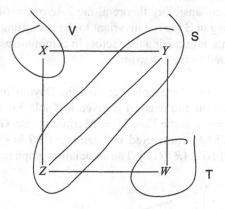

- If a variable is independent of all other variables, then it constitutes a separate component of the association graph, and its categories can be summed over without affecting any of the other λ-terms. In the association graph below, W is independent of X, Y, and Z, and so it may be collapsed over without affecting any of the λ-terms involving X, Y, and Z.

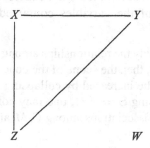

54

- The collapsibility theorem presents sufficient but not necessary conditions for collapsing over the levels of a variable. That is, conditional independence (separation in the association graph) is sufficient but not necessary for collapsibility. See Fienberg (1981, sec. 3.8) and Agresti (2002, p. 398) for further discussion.

- As seen in Example 5.2.3 above, there are often choices in forming the partition S, V, and T, which then determine which variables can be collapsed over and which λ-terms are preserved. Generally, one would make these choices in a way that most effectively addresses the research questions and interests. Often it is advantageous to create the partition in such a way that S contains a minimum of factors. The reason for this is that, according to the collapsibility theorem, the λ-terms involving the factors that are exclusively in S are not invariant to the collapsing of the factors in either V or T. Thus, minimizing the factors in S minimizes the associations that may be affected by the collapsing.

Example 5.3.1. As a final example, consider the Dayton high school survey of Example 4.3.7. The full data set is given in Table 5.6. Recall that $A =$ Alcohol Use, $C =$ Cigarette Use, $M =$ Marijuana Use, $G =$ Gender, and $R =$ Race. The LLM considered in Example 4.3.7 has generating class $[AC][AM][CM][AG][AR][GR]$. The association graph for this LLM is

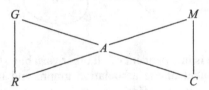

In terms of collapsibility, there are several partitioning choices: $S = \{A\}$, $S = \{A, M\}$, $S = \{A, C\}$, $S = \{A, G\}$, or $S = \{A, R\}$. By selecting $S = \{A\}$, no first-order interactions will be affected by any allowable collapsing, but each of the other choices may lead to a distorted first-order interaction of the variables contained exclusively in S on collapsing.

If it is of interest to study the relationships among Alcohol Use, Cigarette Use, and Marijuana Use, then the scope of the conclusions and the stability of the estimators could be increased by collapsing over Gender and Race. Consequently, by choosing $S = \{A\}$, one may collapse over Gender and Race to safely study the associations among C, M, and A.

Table 5.6. Cross-Classification of 2,276 Nonurban High School Seniors According to Cigarette, Alcohol, and Marijuana Use, and Gender and Ethnic Background

				Gender (G)	
Alcohol Use (A)	Cigarette Use (C)	Marijuana Use (M)	Ethnic (E)	Female	Male
Yes	Yes	Yes	White	405	453
			Other	23	30
		No	White	268	228
			Other	23	19
	No	Yes	White	13	28
			Other	2	1
		No	White	218	201
			Other	19	18
No	Yes	Yes	White	1	1
			Other	0	1
		No	White	17	17
			Other	1	8
	No	Yes	White	1	1
			Other	0	0
		No	White	117	133
			Other	12	17

SOURCE: Wright State University Boonshoft School of Medicine and United Health Services 1992 Survey, Dayton, Ohio.

NOTE: See Example 5.3.1.

The association graph has been used to good effect in analyzing and interpreting a given hierarchical LLM. Specifically, it has been used to determine decomposability, find conditional independencies, and identify collapsibility conditions. It can also be used to achieve closed-form factorization of joint probabilities in decomposable LLMs, but the procedure is quite involved (see, e.g., Khamis & McKee, 1997). In the next chapter, the generator multigraph is introduced as an alternative graphical technique for analyzing and interpreting LLMs.

CHAPTER 6. THE GENERATOR MULTIGRAPH

The *generator multigraph*, or simply *multigraph*, will be constructed for a hierarchical loglinear model (LLM), and then graph theory principles will be used to analyze and interpret the model. The multigraph is a powerful tool that serves as an alternative to the association graph for the purpose of analyzing and interpreting hierarchical LLMs.

6.1 Construction of the Multigraph

A *multigraph* is a mathematical graph in which it is possible to have more than one edge joining two vertices. In the present application, the mathematical graph that will be constructed is one in which the vertices are the generators of the LLM generating class and the edges (or *multiedges*) that join two such generators are equal in number to the number of indices common to the two generators (i.e., the number of indices in the intersection of the two sets of indices corresponding to the two generators being joined). In short, the multigraph for a given LLM has the following:

Vertex set = Generators in the generating class

Multiedge set = Edges equal in number to the number of indices
common to the two generators being joined

6.2 Multigraphs for Three-Way Tables

We now construct the multigraph for each of the five LLMs for the three-way contingency table (see Section 4.2 for the corresponding association graphs).

6.2.1. Mutual Independence: $log(\mu_{ijk}) = \lambda + \lambda_i^X + \lambda_j^Y + \lambda_k^Z$

The generating class is $[X][Y][Z]$. The vertices of the multigraph are the indices corresponding to the generators of the model, X, Y, and Z. Since none of the vertices share an index, there are no edges. Thus, the multigraph for this generating class is

X

Y Z

Note that this multigraph is identical to the association graph in Subsection 4.2.1.

6.2.2. Joint Independence: $log(\mu_{ijk}) = \lambda + \lambda_i^X + \lambda_j^Y + \lambda_k^Z + \lambda_{jk}^{YZ}$

The generating class for this LLM is $[X][YZ]$. The multigraph is

$$X \qquad\qquad\qquad YZ$$

Since the two vertices, X and YZ, share no indices, there are no edges.

6.2.3. Conditional Independence:
$log(\mu_{ijk}) = \lambda + \lambda_i^X + \lambda_j^Y + \lambda_k^Z + \lambda_{ik}^{XZ} + \lambda_{jk}^{YZ}$

The generating class is $[XZ][YZ]$. The multigraph is

$$XZ \text{———————} YZ$$

Since the two vertices share a single index, Z (i.e., $\{X, Z\} \cap \{Y, Z\} = \{Z\}$), there is a single edge joining them.

6.2.4. Homogeneous Association:
$log(\mu_{ijk}) = \lambda + \lambda_i^X + \lambda_j^Y + \lambda_k^Z + \lambda_{ij}^{XY} + \lambda_{ik}^{XZ} + \lambda_{jk}^{YZ}$

The generating class is $[XY][XZ][YZ]$. The multigraph is

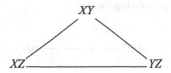

Here, each pair of vertices share a single index, so there is a single edge joining each pair of vertices.

6.2.5. Saturated Model:
$log(\mu_{ijk}) = \lambda + \lambda_i^X + \lambda_j^Y + \lambda_k^Z + \lambda_{ij}^{XY} + \lambda_{ik}^{XZ} + \lambda_{jk}^{YZ} + \lambda_{ijk}^{XYZ}$

The generating class is $[XYZ]$. The multigraph is

$$XYZ$$

Here, the multigraph consists of the single vertex, XYZ.

58

6.3 Multigraphs for Multi-Way Tables

The multigraph will now be constructed for each of Examples 4.3.1 to 4.3.8.

Example 6.3.1. Consider the LLM having generating class [ABC][BD] [CD]. The three generators, [ABC], [BD], and [CD], form the vertices of the multigraph. There is a single index common to [ABC] and [BD], namely, {B}, so a single edge joins these two vertices. Likewise, there is a single index common to [ABC] and [CD], namely, {C}, so there is a single edge joining these two vertices. Finally, there is a single index common to [BD] and [CD], namely, {D}, so there is a single edge joining these two vertices. Then the multigraph representation of the generating class [ABC][BD][CD] is

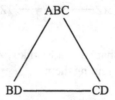

Example 6.3.2. Consider the LLM having generating class [ABC][BD]. Only the index {B} is common to the two generators, so the multigraph representation of this generating class is

ABC———————BD

Example 6.3.3. Consider the generating class [AB][BD][CD][AC]. The multigraph is

Example 6.3.4. Consider the generating class [AS][ACR][MCS][MAC]. The multigraph is

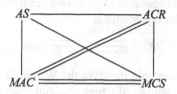

Note here that there are two indices common to [ACR] and [MAC], so there is a double edge joining these two vertices. Similarly, there are two indices common to [MAC] and [MCS], so there is a double edge joining these two vertices.

Example 6.3.5. Consider the generating class [ABCD][ACE][BCG] [CDF]. The multigraph is

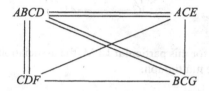

In constructing the multigraph for more complex LLMs such as this, it is often helpful to enumerate the pairs of vertices. In this case, the enumeration of the six pairs of vertices and the indices common to the two vertices in each case is as follows:

Pairs of Vertices	Indices Common to the Two Vertices	Number of Edges
[ABCD], [ACE]	A, C	2
[ABCD], [BCG]	B, C	2
[ABCD], [CDF]	C, D	2
[ACE], [BCG]	C	1
[ACE], [CDF]	C	1
[BCG], [CDF]	C	1

Example 6.3.6. This example uses the Edwards and Kreiner (1983) data from the Institute for Social Research, Copenhagen (see Table 4.1). The LLM that the authors considered has generating class [*ARME*][*AMET*]. There are three indices common to these two generators, namely, *A*, *M*, and *E*, so the multigraph for this model is

$$ARME \equiv\equiv\equiv\equiv AMET$$

Example 6.3.7. This example involves the Wright State University/United Way collaborative study of high school alcohol, cigarette, and marijuana use (see Table 4.2). The best-fitting model has generating class [*AC*][*AM*][*CM*][*AG*][*AR*][*GR*]. The multigraph for this model is

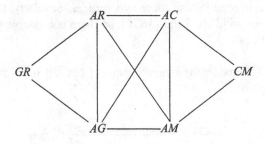

As will be seen for this particular LLM, the association graph is easier to work with than the multigraph.

Example 6.3.8. Consider the generating class (using the numbers 0 through 9 to represent 10 categorical variables) [67][013][125][178][1347][1457] [1479]. The multigraph is

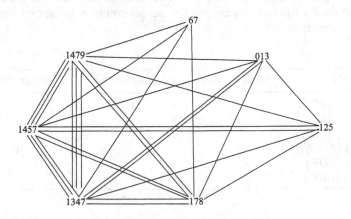

It does not appear that this multigraph is any less complicated than the association graph given in Example 4.3.8 (it may even look *more* complex!). However, as will be seen, the analysis of this multigraph is far easier than the analysis of the association graph in terms of the applications used here.

6.4 Maximum Spanning Trees

A *connected graph* is any graph for which there is at least one path from any vertex to any other vertex in the graph. The graph on the left below is connected, while the one on the right is not connected.

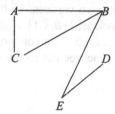 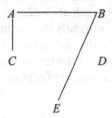

A *tree* is a connected graph with no circuit (or closed loop) that includes each vertex of the graph. The graph on the left above is not a tree because there is a circuit including the vertices A, B, and C. The graph on the right above is not a tree because it is not connected. The graph below is a tree.

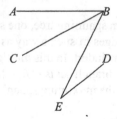

For a multigraph, a *maximum spanning tree* is a tree in which the sum of all the edges is maximum. Maximum spanning trees are not unique; they always exist, and they are typically easy to identify. Kruskal's (1956) algorithm is a technique in which the maximum spanning tree is obtained by successively selecting multiedges with the maximum number of edges so that no circuits are formed and all vertices are included. The family of sets of factor indices (i.e., the indices common to two vertices being joined) in the maximum spanning tree are called the *branches* of the tree. By way of illustration, consider the multigraph below:

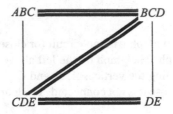

The only possible tree for this multigraph is the double edge joining the vertices ABC and BCD; the indices, B and C (in boldface), included in the multigraph above denote the indices common to the two vertices. So this tree is trivially the maximum spanning tree with branch set $\{\{B, C\}\}$.

For the multigraph below, there are several choices for trees but only one choice for the maximum spanning tree; that is, only one tree has the maximum number of edges.

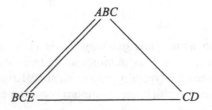

In selecting the maximum spanning tree, one selects the multiedges having the largest number of edges in such a way as to avoid creating a circuit and so that all vertices are included. In this multigraph, the maximum spanning tree is in boldface. The branch set is $\{\{B, C\}, \{C, D\}, \{D, E\}\}$.

In the multigraph below, the maximum spanning tree is not unique.

Certainly the double edge joining ABC and BCE must be included in the maximum spanning tree, but either one of the single edges may be included to complete the tree. Regardless of which single edge is included, note that the set of branches is $\{\{B, C\}, \{C\}\}$.

In the following, a maximum spanning tree will be identified for each of Examples 6.3.1 through 6.3.8.

Example 6.4.1. The generating class is $[ABC][BD][CD]$. The multigraph is

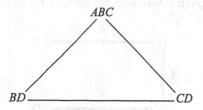

In this case, there are three possible spanning trees:

1. The edge joining $[ABC]$ to $[CD]$ and the edge joining $[CD]$ to $[BD]$

2. The edge joining $[CD]$ to $[BD]$ and the edge joining $[BD]$ to $[ABC]$

3. The edge joining $[ABC]$ to $[BD]$ and the edge joining $[ABC]$ to $[CD]$

Since each of these spanning trees has multiplicity two (i.e., the total number of edges in each case is two), any of the three choices would be a maximum spanning tree. In this case, the maximum spanning tree is not unique. Notationally, we can identify each of the above maximum spanning trees by using the families of indices common to the two vertices being joined, called the branch set: (1) $\{\{C\}, \{D\}\}$; (2) $\{\{D\}, \{B\}\}$; and (3) $\{\{B\}, \{C\}\}$, respectively. In the multigraph below, each edge is labeled with the index common to the two vertices being joined (in boldface), and the edges chosen for the maximum spanning tree are identified by the bold font; in this case, the third choice from above is selected. The branches of this maximum spanning tree are $\{\{B\}, \{C\}\}$.

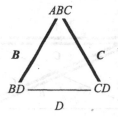

64

Example 6.4.2. The generating class is $[ABC][BD]$. Trivially, the maximum spanning tree is the tree consisting of the single edge (corresponding to the index B) joining $[ABC]$ to $[BD]$.

$$ABC \overset{B}{\rule{3cm}{0.4pt}} BD$$

Example 6.4.3. The generating class is $[AB][BD][CD][AC]$. There are four choices of maximum spanning trees in this case, each of multiplicity three. One possibility is given below, with branches $\{\{A\}, \{B\}, \{C\}\}$:

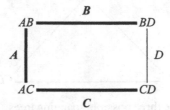

Example 6.4.4. The generating class is $[AS][ACR][MCS][MAC]$. Here, the double edges $\{\{A\}, \{C\}\}$ and $\{\{M\}, \{C\}\}$ must be included in the maximum spanning tree because each is of multiplicity two. To include the vertex $[AS]$, one may use the edge joining it to $[MAC]$, $[ACR]$, or $[MCS]$. Selecting the first of these choices, the branches of the maximum spanning tree are $\{\{A, C\}, \{M, C\}, \{A\}\}$; see the graph below:

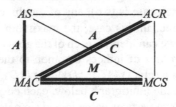

Example 6.4.5. The generating class is $[ABCD][ACE][BCG][CDF]$. Here, there is a unique maximum spanning tree, $\{\{A, C\}, \{B, C,\}, \{C, D\}\}$, involving six edges.

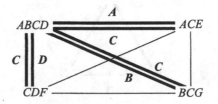

Example 6.4.6. The generating class is [*ARME*][*AMET*]. Trivially, the maximum spanning tree in this case has one branch: {{*A, M, E*}}.

Example 6.4.7. The generating class is [*AC*][*AM*][*CM*][*AG*][*AR*][*GR*]. The multigraph for this model is

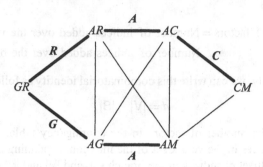

There are many choices for the maximum spanning tree; the one shown above is one possible choice. The branch set is {{*C*}, {*A*}, {*R*}, {*G*}, {*A*}}. Note that the branch set is a multiset because a given element can appear in the set more than once. The multiplicity of this branch set is then five.

Example 6.4.8. The generating class is [67][013][125][178][1347][1457] [1479]. There are many choices for a maximum spanning tree in this multigraph. One such choice is given below, with the branches of the maximum spanning tree shown in boldface. The branches are {{1, 4, 7}, {1, 4, 7}, {1, 7}, {1, 3}, {1, 5}, {7}}.

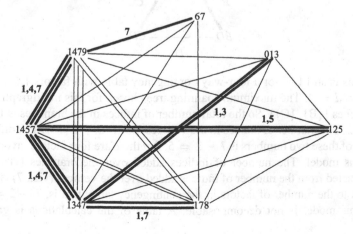

6.5 Decomposability

McKee and Khamis (1996, Theorem 1) show that a hierarchical LLM is decomposable if and only if the number of indices added over the branches in any maximum spanning tree of the multigraph subtracted from the number of indices added over the vertices of the multigraph is equal to the dimensionality of the contingency table. That is, the LLM is decomposable if and only if

Number of factors = Number of indices added over the vertices −

Number of indices added over the branches.

Notationally, we can write this combinatorial identity as follows:

$$d = |V| - |B|, \tag{6.1}$$

where d is the number of factors in the contingency table, V represents the set of indices in the vertex set of the maximum spanning tree, B represents the multiset of indices in the branch set, and $|V|$ and $|B|$ represent the cardinality of V and B, respectively.

Each of Examples 6.4.1 to 6.4.8 will be reviewed, and it will be determined if the model is decomposable. In each case, the multigraph with the maximum spanning tree obtained in Section 6.4 is used.

Example 6.5.1. Generating class: $[ABC][BD][CD]$.

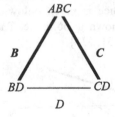

This is an LLM for a four-way contingency table with factors $\{A, B, C, D\}$, so $d = 4$. The maximum spanning tree chosen for this multigraph has branches $\{\{B\}, \{C\}\}$. So the total number of indices in the branches is $1 + 1 = 2$. The number of indices in the vertices is $3 + 2 + 2 = 7$. The difference of these two numbers is $7 - 2 = 5$. But there are four factors involved in this model. The number of indices added over the branches (viz., 2) subtracted from the number of indices added over the vertices (viz., 7) is not equal to the number of factors in the contingency table; that is, $7 - 2 \neq 4$. So this model is not decomposable. A table of the calculations is given below; the two numbers in boldface must agree for the model to be decomposable.

Number of Indices Added Over Vertices	Number of Indices Added Over Branches	Difference	Number of Factors	Decomposable?
7	2	**5**	**4**	No

Example 6.5.2. Generating class: [*ABC*][*BD*].

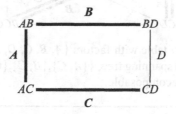

$$ABC \xrightarrow{\hspace{1em} B \hspace{1em}} BD$$

This is a four-way table with factors {*A, B, C, D*}. There is only one branch: {*B*}. So we have

Number of Indices Added Over Vertices	Number of Indices Added Over Branches	Difference	Number of Factors	Decomposable?
5	1	**4**	**4**	Yes

Example 6.5.3. Generating class: [*AB*][*BD*][*CD*][*AC*].

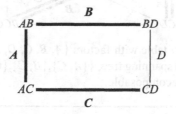

This is a four-way table with factors {*A, B, C, D*}. The branches are {{A}, {B}, {C}}. This LLM is not decomposable.

Number of Indices Added Over Vertices	Number of Indices Added Over Branches	Difference	Number of Factors	Decomposable?
8	3	**5**	**4**	No

Example 6.5.4. Generating class: $[AS][ACR][MCS][MAC]$.

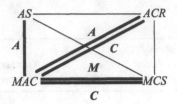

This is a five-way table with factors $\{A, C, M, R, S\}$. We use the maximum spanning tree with branches $\{\{A, C\}, \{C, M\}, \{A\}\}$. This LLM is not decomposable.

Number of Indices Added Over Vertices	Number of Indices Added Over Branches	Difference	Number of Factors	Decomposable?
11	5	6	5	No

Example 6.5.5. Generating class: $[ABCD][ACE][BCG][CDF]$.

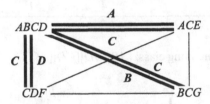

This is a seven-way table with factors $\{A, B, C, D, E, F, G\}$. Here, there is a unique maximum spanning tree, $\{\{A, C\}, \{B, C\}, \{C, D\}\}$, involving six edges. This LLM is decomposable.

Number of Indices Added Over Vertices	Number of Indices Added Over Branches	Difference	Number of Factors	Decomposable?
13	6	7	7	Yes

Example 6.5.6. Generating class: $[ARME][AMET]$.

$$A, M, E$$
$$ARME \equiv\equiv\equiv AMET$$

This is a five-way table with factors $\{A, E, M, R, T\}$. Trivially, the maximum spanning tree in this case has one triple edge with branch set $\{\{A, M, E\}\}$. This LLM is decomposable.

Number of Indices Added Over Vertices	Number of Indices Added Over Branches	Difference	Number of Factors	Decomposable?
8	3	5	5	Yes

Example 6.5.7. Generating class: [AC][AM][CM][AG][AR][GR].

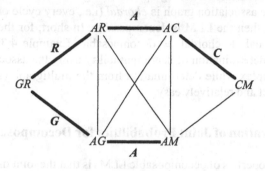

This is a five-way table with factors $\{A, C, G, M, R\}$. The branches of the maximum spanning tree are $\{\{C\}, \{A\}, \{R\}, \{G\}, \{A\}\}$. This LLM is not decomposable.

Number of Indices Added Over Vertices	Number of Indices Added Over Branches	Difference	Number of Factors	Decomposable?
12	5	7	5	No

Example 6.5.8. Generating class: [67][013][125][178][1347][1457][1479]. This is a 10-way table with factors $\{0, 1, 2, 3, 4, 5, 6, 7, 8, 9\}$. Recall from Example 6.4.8 that the branch set is $\{\{1, 4, 7\}, \{1, 4, 7\}, \{1, 7\}, \{1, 3\}, \{1, 5\}, \{7\}\}$. See Subsection 6.4.8 for the multigraph and maximum spanning tree. This LLM is decomposable.

Number of Indices Added Over Vertices	Number of Indices Added Over Branches	Difference	Number of Factors	Decomposable?
23	13	10	10	Yes

Note the ease with which decomposability can be determined in the multigraph, namely, by applying the simple combinatorial identity in Equation 6.1. The LLMs from among Examples 6.5.1 to 6.5.8 that are decomposable are those in Examples 6.5.2, 6.5.5, 6.5.6, and 6.5.8, in agreement with the conclusions made in Examples 4.3.1 to 4.3.8 using the association graph. Recall that, using the association graph, if the LLM is *graphical* (i.e., the maxcliques in the association graph $\equiv$ the generators in the generating class) *and* the association graph is *chordal* (i.e., every cycle of length four has a chord), then the LLM is decomposable. In short, for the association graph, graphical + chordal = decomposable. Example 4.3.8 shows an LLM where determination of decomposability using the association graph is rather complex while determination from the multigraph (see Example 6.5.8) is direct and relatively easy.

6.6 Factorization of Joint Probabilities for Decomposable LLMs

One of the properties of decomposable LLMs is that the joint distribution of the contingency table can be factored in closed form. In fact, the factorization involves marginal probabilities indexed on the indices in the generators of the LLM (i.e., the indices of the marginal probabilities come from the generators of the LLM). This is made more precise as follows.

Consider a contingency table involving d factors. Let $P[\ell_1, \ell_2, \ldots, \ell_d]$ represent the probability that a subject is in level ℓ_1 of the first factor, level ℓ_2 of the second factor, ..., and level ℓ_d of the dth factor. If S is any subset of indices from the set $\{1, 2, \ldots, d\}$, then p_S represents the marginal probability having indices contained in S while all other indices are summed over. As an illustration, consider a four-way table with factors $\{A, B, C, D\}$. Here, A is the first factor, B is the second factor, and so on. If the LLM under consideration is $[ABC][BCD]$, then there is a single branch (double-edged) in the maximum spanning tree: $\{B, C\}$. The multigraph and the maximum spanning tree are given here:

$$ABC \overset{B}{\underset{C}{=\!=\!=\!=\!=}} BCD$$

So $V = \{\{A, B, C\}, \{B, C, D\}\}$ and $\mathbf{B} = \{B, C\}$. Then, if $S = \{B, C\}$, $p_S = p_{+\ell_2\ell_3+}$; this represents the probability in the marginal two-way table of level ℓ_2 of factor B and level ℓ_3 of factor C after collapsing over the levels of factors A and D. Similarly, if $S = \{A, B, C\}$, then $p_S = p_{\ell_1\ell_2\ell_3+}$, the marginal probability for levels ℓ_1, ℓ_2, and ℓ_3 of factors A, B, and C, respectively. (The levels of factor D are summed over or collapsed over.)

McKee and Khamis (1996, Theorem 2) show that for a decomposable model, the joint probability, $P[\ell_1, \ell_2, \ldots, \ell_d]$, can be factored as follows:

$$P[\ell_1, \ell_2, \ldots, \ell_d] = \frac{\displaystyle\prod_{S \in V} p_S}{\displaystyle\prod_{S \in B} p_S}. \tag{6.2}$$

That is, the numerator is the product of marginal probabilities where the indices come from the generators of the LLM (hence, the number of factors in the numerator is equal to the number of generators in the generating class of the model); the denominator is the product of marginal probabilities where the indices come from the branches in a maximum spanning tree (hence, the number of factors in the denominator is equal to the number of branches in the maximum spanning tree of the multigraph). By way of illustration, consider the model above, $[ABC][BCD]$. The maximum spanning tree trivially has a single branch with two edges: $\{B, C\}$. The model is decomposable because $6 - 2 = 4$ and there are four factors in the contingency table. The factorization of the joint probability is

$$P[\ell_1, \ell_2, \ell_3, \ell_4] = \frac{\displaystyle\prod_{S \in \{\{A,B,C\}\{B,C,D\}\}} p_S}{\displaystyle\prod_{S \in \{\{B,C\}\}} p_S} = \frac{p_{\ell_1\ell_2\ell_3+}\, p_{+\ell_2\ell_3\ell_4}}{p_{+\ell_2\ell_3+}}.$$

Note that in the numerator, the indices ℓ_1, ℓ_2, and ℓ_3 for the first factor correspond to the indices in the generator $[ABC]$ and the indices ℓ_2, ℓ_3, and ℓ_4 for the second factor correspond to the indices in the generator $[BCD]$; in the denominator, the indices ℓ_2 and ℓ_3 correspond to the indices in the branch, $\{B, C\}$.

Each of Examples 6.5.1 to 6.5.8 will be considered below, and for those models that are decomposable, the joint probability will be factored in closed form using Equation 6.2 above.

Example 6.6.1. Generating class: $[ABC][BD][CD]$. The LLM is nondecomposable.

Example 6.6.2. Generating class: $[ABC][BD]$. Here, $V = \{\{A, B, C\}, \{B, D\}\}$ and $B = \{B\}$. The joint probability for this model is

$$P[\ell_1, \ell_2, \ell_3, \ell_4] = \frac{\prod\limits_{S \in \{\{A,B,C\}\{B,D\}\}} p_S}{\prod\limits_{S \in \{B\}} p_S} = \frac{p_{\ell_1\ell_2\ell_3 + p + \ell_2 + \ell_4}}{p_{+\ell_2++}}.$$

Example 6.6.3. Generating class: $[AB][BD][CD][AC]$. The LLM is nondecomposable.

Example 6.6.4. Generating class: $[AS][ACR][MCS][MAC]$. The LLM is nondecomposable.

Example 6.6.5. Generating class: $[ABCD][ACE][BCG][CDF]$. This is a seven-way table with factors $\{A, B, C, D, E, F, G\}$. The maximum spanning tree is $\{\{A, C\}, \{B, C,\}, \{C, D\}\}$. Then, $V = \{\{A, B, C, D\}, \{A, C, E\}, \{B, C, G\}, \{C, D, F\}\}$ and $B = \{\{A, C\}, \{B, C,\}, \{C, D\}\}$. The joint probability for this model is

$$P[\ell_1, \ell_2, \ell_3, \ell_4, \ell_5, \ell_6, \ell_7] =$$
$$\frac{p_{\ell_1\ell_2\ell_3\ell_4+++} \, p_{\ell_1+\ell_3+\ell_5++} \, p_{+\ell_2\ell_3+++\ell_7} \, p_{++\ell_3\ell_4+\ell_6+}}{p_{\ell_1+\ell_3++++} \, p_{+\ell_2\ell_3++++} \, p_{++\ell_3\ell_4+++}}.$$

Example 6.6.6. Generating class: $[ARME][AMET]$. This is a five-way table with factors (listed alphabetically) $\{A, E, M, R, T\}$. Here, A, E, M, R, and T are the first, second, third, fourth, and fifth factors, respectively. Trivially, the maximum spanning tree in this case has one branch: $\{\{A, M, E\}\}$. So $V = \{\{A, E, M, R\}, \{A, E, M, T\}\}$ and $B = \{\{A, E, M\}\}$. We have

$$P[\ell_1, \ell_2, \ell_3, \ell_4, \ell_5] = \frac{p_{\ell_1\ell_2\ell_3\ell_4 + p_{\ell_1\ell_2\ell_3+\ell_5}}}{p_{\ell_1\ell_2\ell_3++}}.$$

Example 6.6.7. Generating class: $[AC][AM][CM][AG][AR][GR]$. The LLM is nondecomposable.

Example 6.6.8. Generating class: $[67][013][125][178][1347][1457][1479]$. This is a 10-way table with factors $\{0, 1, 2, 3, 4, 5, 6, 7, 8, 9\}$. Recall from Example 6.5.8 that the branch set is $\{\{1, 4, 7\}, \{1, 4, 7\}, \{1, 7\}, \{1, 3\}, \{1, 5\}, \{7\}\}$. Then the factorization is

$$P[\ell_0, \ell_1, \ell_2, \ell_3, \ell_4, \ell_5, \ell_6, \ell_7, \ell_8, \ell_9]$$
$$= \frac{P+++++++\ell_6\ell_7+++P\ell_0\ell_1+\ell_3++++++P+\ell_1\ell_2+++\ell_5++++P+\ell_1+++++\ell_7\ell_8+P+\ell_1+\ell_3\ell_4++\ell_7++P+\ell_1+++\ell_4\ell_5+\ell_7+++P+\ell_1+++\ell_4++\ell_7+\ell_9}{(P+\ell_1+++\ell_4++\ell_7++)^2 P+\ell_1+++++\ell_7++P+\ell_1+\ell_3+++++++P+\ell_1+++\ell_5++++P+++++++\ell_7++}.$$

Recall that a given generator multigraph need not have a unique maximum spanning tree. Even for decomposable models, the maximum spanning tree need not be unique; see Example 6.5.8, which shows a decomposable model having more than one maximum spanning tree. However, McKee and Khamis (1996, Lemma 2) show that the multiset of branches, B, for a decomposable model is unique. For instance, in Example 6.5.8, regardless of which maximum spanning tree is selected, the multiset of branches is B = {{1, 4, 7}, {1, 4, 7}, {1, 7}, {1, 3}, {1, 5}, {7}}. Because of the uniqueness of the multiset of branches of the maximum spanning tree in decomposable LLMs, the formula for the factorization of the joint probability (see Equation 6.2 above) is well-defined.

6.7 Fundamental Conditional Independencies in Decomposable LLMs

One of the principal goals in studying categorical variables is to determine which factors are independent of each other and which factors are conditionally independent. In the association graph, this was generally easy to do because "separation" in the graph corresponded to conditional independence (see Subsection 4.1.2). However, there may be cases in which the association graph is too complicated to use; see Example 4.3.8.

The multigraph allows for a relatively simple, step-by-step procedure that identifies *all* the conditional independencies for a given LLM. While such a task is easily accomplished for small contingency tables merely by scrutinizing the generating class or working with the association graph, for large, complex LLMs (especially for decomposable models and models with relatively few generators), the multigraph approach is far superior because of its ease, orderly process, and comprehensiveness.

Denote a partition of the d factors in a contingency table by the $k+1$ sets of factors $C_1, C_2, \ldots, C_k$, and S, where $2 \leq k \leq d - 1$. Recall that the notation "$[C_1 \otimes C_2 \otimes \cdots \otimes C_k | S]$" means that the factors contained in C_1, $C_2, \ldots, C_k$ are mutually independent of each other conditional on S. McKee and Khamis (1996, Theorem 3) show that the generating class for a given LLM uniquely determines a set of *fundamental conditional independencies* (FCIs), each of the form

$$[C_1 \otimes C_2 \otimes \cdots \otimes C_k | S],$$

such that all other conditional independencies can be deduced from these FCIs by replacing S with S' such that $S \subseteq S'$, replacing each C_i with C'_i such that $C'_i \subseteq C_i$, subject to

$$(C'_1 \cup C'_2 \cup \cdots \cup C'_k) \cap S' = \varnothing,$$

and forming appropriate conjunctions. That is, suppose for a given LLM we have the conditional independence relation $[A, B \otimes D|E]$. Then the following conditional independence relations are also true: $[A \otimes D|B, E]$ and $[B \otimes D|A, E]$.

To derive these FCIs, we do the following. Denote the multigraph by M. Suppose there are d factors in the contingency table. Let S be a subset of these factors. Now construct a new multigraph M/S (read "M-mod-S") by removing each factor of S from each generator (vertex in the multigraph) and removing each edge corresponding to that factor. Then the FCI corresponds to the mutual independence of the sets of factors in the disconnected components of M/S conditional on S.

For the case of decomposable models, S is chosen to be the branches of the maximum spanning tree. Suppose we wish to analyze the model of conditional independence in the three-way contingency table, $[XZ][YZ]$ (this is a decomposable model). The vertex set and branch set are $V = \{\{X, Z\}, \{Y, Z\}\}$ and $B = \{\{Z\}\}$, respectively. The multigraph, M, and the maximum spanning tree are

$$XZ \underset{}{\overset{Z}{\rule{3cm}{0.4pt}}} YZ$$

Choose S to be the branch in the maximum spanning tree; that is, $S = \{Z\}$. To construct M/S, remove the index Z from each vertex, and remove the edge corresponding to Z:

$$X \qquad \qquad Y$$

Thus, M/S results in two disconnected components, $\{X\}$ and $\{Y\}$. In this case, $C_1 = \{X\}$ and $C_2 = \{Y\}$. The FCI is of the form $[C_1 \otimes C_2|S]$; that is, $[X \otimes Y|Z]$. So X and Y are independent conditional on Z. This interpretation agrees with the discussion in Subsections 2.3.3 (probabilistic model) and 4.2.3 (association graph). Now, FCIs will be determined for each of Examples 6.5.1 to 6.5.8 where the model is decomposable. The case of non-decomposable models will be addressed in the next chapter.

Example 6.7.1. Generating class: $[ABC][BD][CD]$. This model is nondecomposable. See Chapter 7 to determine how FCIs are identified.

Example 6.7.2. Generating class: $[ABC][BD]$. The multigraph, M, is given below:

$$ABC \underset{}{\overset{B}{\rule{3cm}{0.4pt}}} BD$$

$V = \{\{A, B, C\}, \{B, D\}\}$ and $B = \{B\}$. Let $S = \{B\}$. Then M/S is

$$AC \qquad\qquad D$$

Then $C_1 = \{A, C\}$, $C_2 = \{D\}$, and $S = \{B\}$. The FCI has the form $[C_1 \otimes C_2|S]$, thus $[A, C \otimes D|B]$; that is, factors A and C are independent of D conditional on B. From this FCI, we can generate another conditional independence relation by reducing C_1 and expanding S: namely, replacing $C_1 = \{A, C\}$ with $C_1' = \{A\}$ and replacing $S = \{B\}$ with $S' = \{B, C\}$. Note that $\left(C_1' \cup C_2\right) \cap S' = \emptyset$. We have $[A \otimes D|B, C]$. Similarly, we have $[C \otimes D|A, B]$.

Example 6.7.3. Generating class: $[AB][BD][CD][AC]$. This model is non-decomposable; see Chapter 7.

Example 6.7.4. Generating class: $[AS][ACR][MCS][MAC]$. This model is nondecomposable; see Chapter 7.

Example 6.7.5. Generating class: $[ABCD][ACE][BCG][CDF]$. The multi-graph, M, is below:

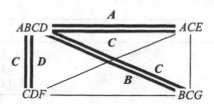

$V = \{\{A, B, C, D\}, \{A, C, E\}, \{B, C, G\}, \{C, D, F\}\}$ and $B = \{\{A, C\}, \{B, C\}, \{C, D\}\}$. Here, we have three choices for S: $S_1 = \{A, C\}$, $S_2 = \{B, C\}$, and $S_3 = \{C, D\}$. Each choice will generate an FCI. For the first FCI, consider M/S_1:

M/S_1 is a multigraph with two disconnected components: $C_1 = \{B, D, F, G\}$ and $C_2 = \{E\}$. The FCI is $[B, D, F, G \otimes E|A, C]$. Numerous other conditional independencies can be generated by reducing C_1 and expanding S_1 according to the construction described above.

For the second FCI, consider M/S_2:

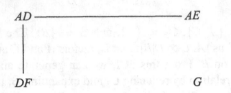

Here, we have $[A, D, E, F \otimes G|B, C]$.

For the third FCI, consider M/S_3:

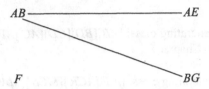

Here, we have $[A, B, E, G \otimes F|C, D]$.

A summary of the interpretation of the LLM having generating class $[ABCD][ACE][BCG][CDF]$ is given below:

S	FCI
$\{A, C\}$	$[B, D, F, G \otimes E|A, C]$
$\{B, C\}$	$[A, D, E, F \otimes G|B, C]$
$\{C, D\}$	$[A, B, E, G \otimes F|C, D]$

There is some redundancy in the conditional independencies that can be generated from these FCIs. For instance, from the first FCI we have $[G \otimes E|A, B, C, D, F]$; but the second FCI can generate the same conditional independence relation.

Example 6.7.6. Generating class: $[ARME][AMET]$. The multigraph, M, is below:

$$A, M, E$$

ARME ≡≡≡ AMET

$V = \{\{A, R, M, E\}, \{A, M, E, T\}\}$ and $B = \{\{A, M, E\}\}$. Let $S = \{A, M, E\}$. Then M/S is

$$R \qquad T$$

So $[R \otimes T|A, M, E]$.

Example 6.7.7. Generating class: $[AC][AM][CM][AG][AR][GR]$. This model is nondecomposable; see chapter 7.

Example 6.7.8. Generating class: $[67][013][125][178][1347][1457][1479]$. Here,

$V = \{\{6,7\}, \{0,1,3\}, \{1,2,5\}, \{1,7,8\}, \{1,3,4,7\}, \{1,4,5,7\}, \{1,4,7,9\}\}$
$B = \{\{1,4,7\}, \{1,4,7\}, \{1,7\}, \{1,3\}, \{1,5\}, \{7\}\}$.

The FCIs are

S	FCI
$\{1, 4, 7\}$	$[0, 3 \otimes 2, 5 \otimes 6 \otimes 8 \otimes 9 | 1, 4, 7]$
$\{1, 7\}$	$[0, 2, 3, 4, 5, 9 \otimes 6 \otimes 8 | 1, 7]$
$\{1, 3\}$	$[2, 4, 5, 6, 7, 8, 9 \otimes 0 | 1, 3]$
$\{1, 5\}$	$[0, 3, 4, 6, 7, 8, 9 \otimes 2 | 1, 5]$
$\{7\}$	$[0, 1, 2, 3, 4, 5, 8, 9 \otimes 6 | 7]$

For illustration purposes, consider the first FCI, where $S = \{1, 4, 7\}$. Then remove the indices 1, 4, and 7 from each vertex, and remove all the edges corresponding to these indices to obtain M/S as follows:

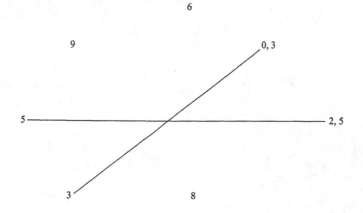

Thus, there are five disconnected components: $C_1 = \{0, 3\}$; $C_2 = \{2, 5\}$; $C_3 = \{6\}$; $C_4 = \{8\}$; and $C_5 = \{9\}$. Then the FCI is of the form $[C_1 \otimes C_2 \otimes C_3 \otimes C_4 \otimes C_5 | S]$, or $[0, 3 \otimes 2, 5 \otimes 6 \otimes 8 \otimes 9 | 1, 4, 7]$. A similar construction of M/S is used to obtain the other FCIs.

The construction of FCIs for decomposable LLMs has been shown above. In this case, S is chosen to be a branch of the maximum spanning tree. For nondecomposable LLMs, a different construction is required based on the choice of S. This is discussed in the next chapter.

CHAPTER 7. FUNDAMENTAL CONDITIONAL INDEPENDENCIES FOR NONDECOMPOSABLE LOGLINEAR MODELS

For nondecomposable loglinear models (LLMs), the construction of fundamental conditional independencies (FCIs) is more complex, partly because the multiset of branches of a maximum spanning tree need not be unique. So we will use a graph theory tool called edge cutsets.

7.1 Edge Cutsets

An *edge cutset* of a multigraph is an inclusion-minimal set of multiedges whose removal disconnects the multigraph. Consider once again the model of conditional independence in the three-way table, $[XZ][YZ]$, having multigraph

$$XZ \underset{}{\overset{Z}{\rule{3cm}{0.4pt}}} YZ$$

In this case, there is a single edge (corresponding to the factor Z) whose removal disconnects the two vertices, and it is trivially the minimum number of edges that does so. So the edge cutset is $\{Z\}$, the factor index associated with the edge whose removal disconnects the multigraph. A convenient way of keeping track of edge cutsets is to draw dotted lines that disconnect the multigraph; those edges that the dotted lines intersect are contained in an edge cutset. For the above example, we have

Note that $[XZ][YZ]$ is a decomposable model. Note also that the edge cutset, $\{Z\}$, is the branch of the maximum spanning tree of the multigraph. It is always the case that for decomposable models, the edge cutsets are the branches of the maximum spanning tree. Check this for the next two multigraphs, which also correspond to decomposable LLMs.

80

For the multigraph below, there's a single edge cutset: $\{B, C, D\}$.

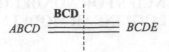

In the multigraph below, there are two edge cutsets, $\{B, C\}$ and $\{C\}$, resulting from the three dotted lines.

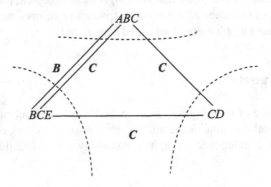

The edge cutsets for each of the nondecomposable LLMs in Examples 6.7.1 to 6.7.8 will now be obtained. As mentioned above, the edge cutsets for the decomposable LLMs are simply the branches of the maximum spanning tree, **S**.

Example 7.1.1. Consider the LLM having generating class $[ABC][BD]$ $[CD]$. Here, there are three edge cutsets (the dotted lines are numbered to identify them; see below): $S_1 = \{B, C\}$, $S_2 = \{C, D\}$, and $S_3 = \{B, D\}$. Consider S_1: The two single edges joining $[ABC]$ to $[BD]$ and $[CD]$ disconnect the multigraph on removal. The index common to $[ABC]$ and $[BD]$ is $\{B\}$, and the index common to $[ABC]$ and $[CD]$ is $\{C\}$. So the edge cutset is $S_1 = \{B, C\}$, and similarly for $S_2 = \{C, D\}$ and $S_3 = \{B, D\}$.

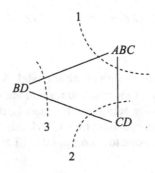

Example 7.1.2. The generating class [*ABC*][*BD*] corresponds to a decomposable model. Thus, its edge cutsets are the branches of the maximum spanning tree. See Example 6.7.2.

Example 7.1.3. Consider the generating class [*AB*][*BD*][*CD*][*AC*]. The multigraph is

For this nondecomposable model, there are six edge cutsets: There are four edge cutsets that each disconnect a single vertex (these are denoted by the dotted lines 1–4 in the figure below), there is an edge cutset that intersects the two horizontal edges (dotted line 5), and there is an edge cutset that intersects the two vertical edges (dotted line 6).

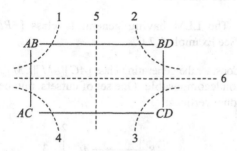

So the two edges associated with the vertex [*AB*] correspond to the indices {*B*} and {*A*} since these are the indices that are common to the two vertices being joined by those edges; removal of these two edges would disconnect this vertex from the rest of the multigraph—see the dotted line 1. Then, $S_1 = \{A, B\}$. Similarly, $S_2 = \{B, D\}$; $S_3 = \{C, D\}$; and $S_4 = \{A, C\}$. Removal of the two horizontal edges would disconnect the vertices [*AB*] and [*AC*] from the vertices [*BD*] and [*CD*]—see the dotted line 5. These two edges correspond to the indices {*B*} and {*C*}. Thus, $S_5 = \{B, C\}$, and similarly for the two vertical edges, where $S_6 = \{A, D\}$.

Example 7.1.4. Consider the generating class [*AS*][*ACR*][*MCS*][*MAC*]. This model is nondecomposable. There are six edge cutsets corresponding to dotted lines having the same orientation as in Example 7.1.3: four edge

82

cutsets disconnecting a single vertex from the rest of the multigraph (dotted lines 1–4) and two edge cutsets disconnecting one pair of vertices from another pair of vertices (dotted lines 5 and 6).

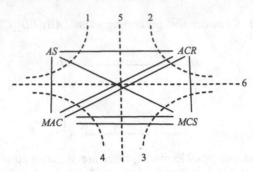

The edge cutsets are $S_1 = \{A, S\}$; $S_2 = \{A, C\}$; $S_3 = \{C, M, S\}$; $S_4 = \{A, C, M\}$; $S_5 = \{A, C, M, S\}$; and $S_6 = \{A, C, S\}$.

Example 7.1.5. The LLM having generating class $[ABCD][ACE][BCG][CDF]$ is decomposable. See Example 6.7.5.

Example 7.1.6. The LLM having generating class $[ARME][AMET]$ is decomposable. See Example 6.7.6.

Example 7.1.7. Consider the generating class $[AC][AM][CM][AG][AR][GR]$. This model is nondecomposable. One set of cutsets, numbered 1 to 6, disconnects individual vertices:

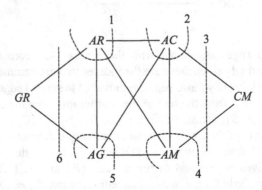

These cutsets are $S_1 = \{A, R\}$; $S_2 = \{A, C\}$; $S_3 = \{C, M\}$; $S_4 = \{A, M\}$; $S_5 = \{A, G\}$; and $S_6 = \{G, R\}$. Another set of cutsets, numbered 7 to 9, corresponds to the vertical dotted lines:

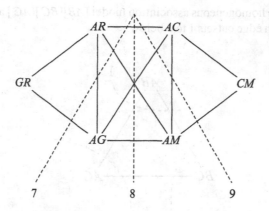

These cutsets are $S_7 = \{A, G\}$; $S_8 = \{A\}$; and $S_9 = \{A, M\}$. Finally, the cutsets numbered 10 to 13 correspond to the horizontal dotted lines:

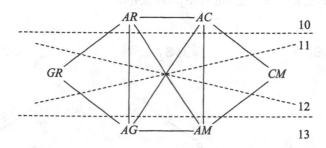

These cutsets are $S_{10} = \{A, C, R\}$; $S_{11} = \{A, C, G\}$; $S_{12} = \{A, M, R\}$; and $S_{13} = \{A, G, M\}$.

Example 7.1.8. The LLM having generating class [67][013][125][178] [1347][1457][1479] is decomposable. See Example 6.7.8.

7.2 FCIs for Nondecomposable LLMs

The identification of FCIs for nondecomposable LLMs is identical to that for decomposable LLMs except that the indices contained in S come from the edge cutsets instead of from the branches of a maximum spanning tree.

Consider the homogeneous association model [AB][BC][AC] and its multi-graph M with edge cutsets 1 to 3 below:

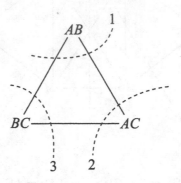

Note that $S_1 = \{A, B\}$; $S_2 = \{A, C\}$; and $S_3 = \{B, C\}$. The multigraphs M/S_i for $i = 1, 2$, and 3 are given below:

M/S_1 M/S_2 M/S_3

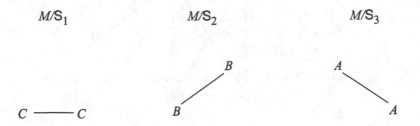

In each case, there is only a single component and, therefore, no conditional independence relation (recall that conditional independencies derive from separate components in the multigraph M/S). Thus, we've confirmed that the homogeneous association model admits no conditional independence among the three factors.

The nondecomposable LLMs from Examples 7.1.1 to 7.1.8 will be used as illustrations of the construction of FCIs.

Example 7.2.1. M has generating class [ABC][BD][CD]; the edge cutsets are $S_1 = \{B, C\}$; $S_2 = \{C, D\}$; and $S_3 = \{B, D\}$ (see Example 7.1.1). The multigraph M/S for each edge cutset is given below:

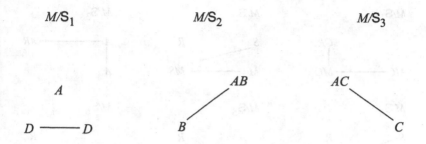

The latter two multigraphs render no FCIs since there is only a single component in each case. The first multigraph, M/S_1, gives the FCI $[A \otimes D \,|\, B, C]$.

Example 7.2.2. The LLM having generating class $[ABC][BD]$ is decomposable; see Example 6.7.2.

Example 7.2.3. M has generating class $[AB][BD][CD][AC]$; the edge cutsets are $S_1 = \{A, B\}$; $S_2 = \{B, D\}$; $S_3 = \{C, D\}$; $S_4 = \{A, C\}$; $S_5 = \{B, C\}$; and $S_6 = \{A, D\}$. Then, M/S_i for $i = 1, 2, \ldots, 6$ are given below:

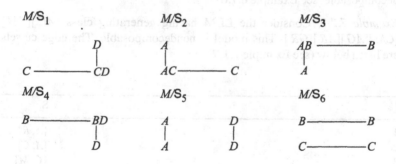

The only multigraphs that yield FCIs are the last two, M/S_5 and M/S_6—namely, $[A \otimes D | B, C]$ and $[B \otimes C | A, D]$, respectively.

Example 7.2.4. M has generating class $[AS][ACR][MCS][MAC]$; the edge cutsets are $S_1 = \{A, S\}$; $S_2 = \{A, C\}$; $S_3 = \{C, M, S\}$; $S_4 = \{A, C, M\}$; $S_5 = \{A, C, M, S\}$; and $S_6 = \{A, C, S\}$. The multigraphs M/S_i for $i = 1, 2, \ldots, 6$ are given below:

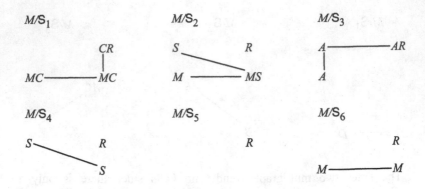

Only M/S_2, M/S_4, and M/S_6 yield FCIs: $[R \otimes M, S | A, C]$; $[R \otimes S | A, C, M]$; and $[R \otimes M | A, C, S]$, respectively. Note that the latter two FCIs are obtainable from the first, $[R \otimes M, S | A, C]$, by using the construction discussed in Section 6.7.

Example 7.2.5. The LLM having generating class $[ABCD][ACE][BCG]$ $[CDF]$ is decomposable; see Example 6.7.5.

Example 7.2.6. The LLM having generating class $[ARME][AMET]$ is decomposable; see Example 6.7.6.

Example 7.2.7. Consider the LLM having generating class $[AC][AM]$ $[CM][AG][AR][GR]$. This model is nondecomposable. The edge cutsets are listed below (see Example 7.1.7):

i	S_i
1	$\{A, R\}$
2	$\{A, C\}$
3	$\{C, M\}$
4	$\{A, M\}$
5	$\{A, G\}$
6	$\{G, R\}$
7	$\{A, G\}$
8	$\{A\}$
9	$\{A, M\}$
10	$\{A, C, R\}$
11	$\{A, C, G\}$
12	$\{A, M, R\}$
13	$\{A, G, M\}$

The nonredundant edge cutsets are S_1, S_2, S_3, S_4, S_5, S_6, S_8, S_{10}, S_{11}, S_{12}, and S_{13}. The multigraphs M/S and the FCIs are given below:

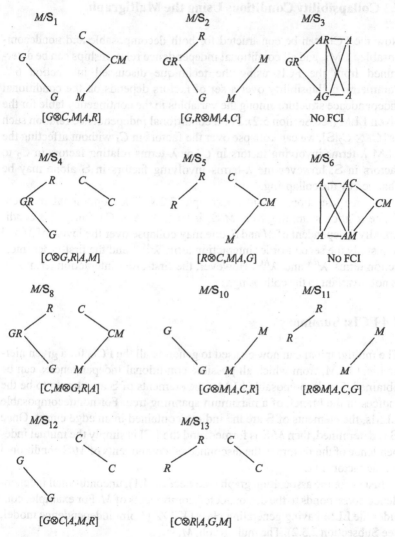

Note that many of the FCIs above are obtainable from other FCIs by using the construction given in Section 6.7. Also note that a much easier way of finding all the conditional independencies for this particular model is to use the association graph and construct partitions S, T, and V, where S separates T and V (see Example 5.3.1).

88

Example 7.2.8. The LLM having generating class [67][013][125][178] [1347][1457][1479] is decomposable; see Example 6.7.8.

7.3 Collapsibility Conditions Using the Multigraph

Now the FCIs can be constructed for both decomposable and nondecomposable LLMs, and all conditional independence relationships can be determined from the FCIs using the technique discussed in Section 6.7. Parametric collapsibility over a set of factors depends on the conditional independence structure among the variables in the contingency table for the given LLM (see Section 5.2). For a conditional independence relation such as $[C_1 \otimes C_2 | S]$, we can collapse over the factors in C_1 without affecting the LLM λ-terms involving factors in C_2 or λ-terms relating factors in C_2 to factors in S; however the λ-terms involving factors in S alone may be changed by the collapsing.

As an illustration, consider Example 7.2.4. The conditional independence relation stemming from M/S_2 is $[R \otimes M, S | A, C]$. Since R is conditionally independent of M and S, one may collapse over the levels of M and S to study the second-order interaction term λ^{ACR} and the first-order interaction terms λ^{RA} and λ^{RC}. However, the first-order interaction term λ^{AC} is not invariant to the collapsing.

7.4 FCIs: Summary

The multigraph M can now be used to generate all the FCIs for a given hierarchical LLM, from which all possible conditional independencies can be obtained. For decomposable LLMs, the elements of S are chosen to be the indices in the branch of a maximum spanning tree. For nondecomposable LLMs, the elements of S are the indices contained in an edge cutset. Once S is determined, then M/S is formed, and the FCI is simply the mutual independence of the factors in the disconnected components of M/S conditional on the factors in S.

Just as for the association graph (see Section 4.1), unconditional independence corresponds to the disconnected components of M. For example, consider the LLM having generating class $[X][YZ]$ (joint independence model; see Subsection 2.3.2). The multigraph, M, is

$$X \qquad YZ$$

The factors in the disconnected components of M are jointly independent; thus, we have $[X \otimes Y, Z]$.

Again, the techniques described in Sections 6.5 (determining decomposability) and 6.6 (factorization of the joint probability) are useful to statistical methodologists and researchers, while those in Section 6.7 and Chapters 5 and 7 (finding FCIs and collapsibility) are of value to the applied researcher and data analyst.

In the next chapter, conclusions and additional examples will be given.

CHAPTER 8. CONCLUSIONS AND
ADDITIONAL EXAMPLES

The association graph and the generator multigraph have been shown to be very useful mathematical tools for analyzing and interpreting hierarchical loglinear models (LLMs). Either or both graphs can be used in any given instance. In some cases the association graph may be easier to use (compare Examples 7.2.7 and 5.3.1), while in other cases the multigraph may be more efficient (compare Examples 6.7.8 and 4.3.8). Taken together, these graphical techniques constitute a useful addition to the collection of statistical methodology procedures for analyzing and interpreting large, complex contingency tables. In this chapter, we will compare and contrast the two approaches and present additional examples to illustrate the techniques that have been presented.

8.1 Comparison of the Association Graph and the Multigraph

Generally, the association graph and the multigraph are equivalent approaches used to accomplish the same things, and in small contingency tables (say $d < 5$) they will do so with comparable amounts of effort. For larger contingency tables, one approach may have an advantage over the other, depending on the desired goal. Let us review the goals in analyzing and interpreting an LLM and compare the two procedures.

8.1.1. Construct the Graph

When the contingency table under consideration is large (i.e., large d) and the LLM of interest has few generators, then the multigraph will be smaller and easier to work with than the association graph (see Example 8.2.3 below). Otherwise, the association graph may be the preferred approach (see Example 8.2.2 below). The association graph is based on the factors in the contingency table (vertices = factors), while the multigraph is based on the generators of the LLM (vertices = generators).

8.1.2. Determine Decomposability

For the association graph, this amounts to determining if the LLM is graphical and the graph is chordal (graphical + chordal = decomposable). This can be quite difficult for large, complicated association graphs (e.g., see Example 4.3.8). For the multigraph, one must verify a combinatorial identity based on a maximum spanning tree; that is, a model is decomposable if and only if $d = |V| - |B|$.

In almost all cases, it is easier to determine decomposability using the multigraph.

8.1.3. Factor the Joint Distribution for Decomposable LLMs

As discussed in Section 4.4, there have been several approaches to factorization of the joint distribution of a decomposable LLM. The association graph approach uses graph neighborhoods and elimination schemes (Darroch et al., 1980; see also Khamis & McKee, 1997, sec. 2). The multigraph approach, based on maximum spanning trees, is comparatively easy (see Section 6.6, Equation 6.2).

8.1.4. Identify all the Conditional Independencies

Using the association graph, one finds all the partitions of the factors of the contingency table, S, T, and V, such that S separates T and V in the graph. For any such given partition, one concludes that the factors in T and V are independent conditional on the factors in S. Finding all such partitions amounts to, in graph-theoretic terms, finding all the "minimal vertex separators" (see Golumbic, 1980). Finding all the minimal vertex separators can be a difficult graph-theoretic task in general (McKee & Khamis, 1996, sec. 4). The multigraph approach uses edge cutsets. Finding all the edge cutsets in the multigraph can also be computationally difficult in large graphs. In fact, in each case there may be an exponential number. However, there is a conceptually simple algorithmic procedure, well-known to electrical engineers, for finding all the edge cutsets in a multigraph (see Gibbons, 1985; McKee & Khamis, 1996, sec. 4; Wilson, 1985). For decomposable LLMs, identification of all the edge cutsets is relatively easy because it is based on a maximum spanning tree. Example 7.2.7 shows an LLM where the association graph is easier to use to find all the conditional independencies (compare with Example 4.3.7); Example 6.7.8 shows an LLM where the multigraph is easier to use (compare with Example 4.3.8).

8.1.5. Collapsibility Conditions

Collapsibility conditions are based on conditional independence relationships in the LLM. Such relationships are determined by (1) the separation of the vertices in the association graph and (2) the FCIs in the multigraph. The ease with which collapsibility conditions are determined depends on the complexity of the graph. For some LLMs the association graph is smaller and easier to work with (i.e., finding partitions S, T, and V), while for other LLMs the multigraph is simpler to work with (i.e., determining the FCIs).

In summary, the association graph and the multigraph provide two different ways to approach the analysis and interpretation of a given contingency table and a given LLM. In addition to the comparisons given above, there may be additional psychological or pedagogical advantages to one over the other within a specific application. Having both approaches available equips the researcher with two different useful and versatile techniques for the analysis and interpretation of hierarchical LLMs. A summary of the comparisons is given in Table 8.1.

Table 8.1. Summary Comparison of the Association Graph and the Multigraph

Goal	Association Graph	Multigraph, M
Graph construction	Vertices = factors Edges = first-order interactions	Vertices = generators Multi-edges = k edges, where k = number of indices common to the two vertices being joined
Determine decomposability	Chordal graph + graphical loglinear model	$d = |V| - |B|$
Factorization of joint probabilities	Elimination schemes	$P[\ell_1, \ell_2, \ldots, \ell_d] = \dfrac{\prod_{S \in V} p_S}{\prod_{S \in B} p_S}$
Find conditional independencies	Partitioning the vertices into subsets S, T, and V	Formation of M/S based on edge cutsets
Interpret collapsibility conditions	Separation of sets of Factors T and V by S determines collapsibility	Based on conditional independencies determined by S and the components of M/S

Example 8.1.1. Consider the LLM having generating class [*ABC*][*BCDE*] [*CDEF*][*CDEG*][*CGHI*] for nine variables $A, B, \ldots, I$. Can you tell, reliably, if this model is decomposable merely by scrutinizing the generating class? Can you identify, reliably, *all* the conditional independencies? Can you determine, reliably, *all* the collapsibility conditions? The last two questions are particularly important for the purpose of interpreting the data and making conclusions about the relationships among the factors. Let us consider this model using the graphical tools that have been presented. Because of the complexity of the association graph in this particular case, we will use the multigraph. The multigraph, M, with a maximum spanning tree (in boldface), is given below:

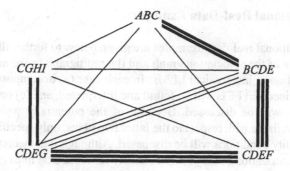

The calculations below show that this model is decomposable:

Number of Indices Added Over Vertices	Number of Indices Added Over Branches	Difference	Number of Factors	Decomposable?
19	10	9	9	Yes

To identify the conditional independencies, we choose S to be the branches of the maximum spanning tree, so there are three distinct choices for S. The resulting multigraphs M/S_i, $i = 1, 2, 3$ are given below along with the FCIs:

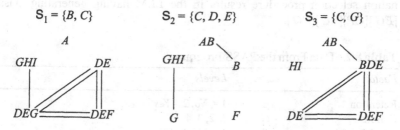

$$S_1 = \{B, C\} \qquad\qquad S_2 = \{C, D, E\} \qquad\qquad S_3 = \{C, G\}$$

$$[A \otimes D, E, F, G, H, I | B, C] \qquad [A, B \otimes F \otimes G, H, I | C, D, E] \qquad [A, B, D, E, F \otimes H, I | C, G]$$

Additional conditional independencies deriving from these FCIs can be obtained by using the construction developed in Section 6.7. Collapsibility conditions are determined from the FCIs directly. For example, corresponding to $S_1 = \{B, C\}$, we have that D, E, F, G, H, and I can be collapsed over without distorting the AB, AC, or ABC interaction terms; the BC interaction term may, however, be distorted.

In the following section, additional real-data examples are presented to provide practice in using the association graph and the multigraph.

94

8.2 Additional Real-Data Examples

Some additional real-data examples are given below to further illustrate the application of the association graph and the multigraph to the analysis and interpretation of hierarchical LLMs. In each case (a) decomposability will be determined, (b) FCIs will be found and interpreted, and (c) collapsibility conditions will be discussed. Because of the potentially wide range of choices available with respect to the latter two items, only specific FCIs and collapsibility conditions will be discussed. Although it is theoretically useful, the factorization of the joint probability for those examples in which the LLM is decomposable will not be provided because of its lack of practical usefulness in interpreting the data.

Example 8.2.1: Wright State University PASS Program

In an effort to improve retention, the University College at Wright State University initiated an academic intervention program, PASS (Preparing for Academic Success Seminar), aimed at students whose GPA was below 2.0 at the end of their first quarter. The pertinent variables and data are given in Table 8.2. Notationally, we have C = Cohort, E = Ethnic Group, G = Gender, P = PASS Participation, and R = Retention. While these data are conducive to sophisticated models (e.g., path models), the client specifically asked that the structural relationships among all the variables be identified using a conservative approach (α = .10). A backward elimination selection procedure results in the LLM having generating class $[EG][CP][RC][PG]$.

Table 8.2. Data From the PASS Program

Factor	Label	Levels
Retention	R	$1 = No, 2 = Yes$
Cohort	C	$1, 2, 3, 4$
PASS Participation	P	$1 = No, 2 = Yes$
Ethnic Group	E	$1 = Caucasian, 2 = African American, 3 = Other$
Gender	G	$1 = Male, 2 = Female$

SOURCE: Data were kindly provided by Dr. Anita Curry-Jackson, Dean, University College, Wright State University.

NOTES: See Example 8.2.1. The cell frequencies, listed lexicographically, with the levels of Gender changing the fastest, the levels of Ethnic Group changing next fastest, . . . , and the levels of Retention changing the slowest, are 9, 4, 5, 4, 0, 4, 57, 46, 18, 29, 6, 15, 3, 7, 1, 4, 1, 1, 12, 8, 1, 10, 1, 1, 22, 20, 3, 10, 3, 3, 10, 8, 8, 10, 1, 2, 12, 10, 2, 8, 0, 1, 3, 6, 2, 3, 0, 1, 4, 6, 4, 6, 0, 1, 57, 48, 12, 26, 5, 14, 9, 5, 6, 5, 0, 3, 22, 21, 6, 16, 1, 1, 39, 18, 21, 15, 2, 5, 19, 25, 6, 22, 1, 2, 25, 22, 4, 18, 4, 1, 15, 19, 3, 9, 1, 3.

The association graph is

$$E \text{————} G \text{————} P \text{————} C \text{————} R$$

The multigraph, M, with maximum spanning tree is

$$EG \text{————} PG$$
$$RC \text{————} CP$$

Decomposability. This LLM is decomposable:

Number of Indices Added Over Vertices	Number of Indices Added Over Branches	Difference	Number of Factors	Decomposable?
8	3	5	5	Yes

FCIs and Interpretation. The multigraphs, M/S, and FCIs for each S are

S = {G}

$$E \quad P$$
$$RC \text{——} CP$$

$[C, P, R \otimes E|G]$

S = {P}

$$EG \text{——} G$$
$$RC \text{——} C$$

$[E, G \otimes R, C|P]$

S = {C}

$$EG \text{——} PG$$
$$R \quad P$$

$[E, G, P \otimes R|C]$

Two variables of particular interest in this study are P (PASS Program) and R (Retention). The third FCI above implies that $[P \otimes R|C, E, G]$; that is, P is independent of R for each cohort, ethnic group, and gender. There is no strong evidence in these data to support a relationship between the PASS program and retention rates. This was disappointing news to the client; however, the program is new and ongoing, and there is hope that with more time, PASS will be seen to increase retention rates.

Collapsibility. An implication of both the second and the third FCIs above is that $[E, G \otimes R|P, C]$ (see Section 6.7). Consequently, one may collapse over E and G and safely analyze the R-P interaction, thereby obtaining, for example, a more stable estimate of the odds ratio or λ-term for the relationship of interest.

Example 8.2.2: Dayton High School Survey

The data for the Dayton High School Survey are given in Table 5.6. In this example, G = Gender, R = Race, A = Alcohol Use, C = Cigarette Use, and M = Marijuana Use. The LLM having generating class $[AC][AM][CM][AG][AR][GR]$ is analyzed in Examples 6.3.7, 6.4.7, 6.5.7, 6.6.7, and 7.1.7 using the multigraph. An alternative model that fits well is $[AC][AM][CM][AG][AR][GM][GR]$ (see Agresti, 2002, pp. 362–363), where the first-order interaction λ^{GM} is added. The multigraph is given below:

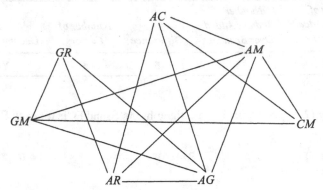

In this case, the association graph would be easier to use; the high number of generators in the generating class produces a rather large, complicated multigraph. The association graph for this generating class is

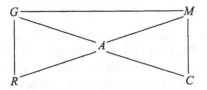

Decomposability. Since $< ARG >$ is a maxclique in this association graph but $[ARG]$ is not a generator in the generating class, this LLM is not graphical and, hence, not decomposable.

FCIs and Interpretation. In identifying conditional independencies, the choices for S would be (1) $\{A, G\}$; (2) $\{A, M\}$; (3) $\{A, G, M\}$; (4) $\{A, M, R\}$; or (5) $\{A, C, G\}$. This leads to five conditional independencies: (1) $[R \otimes C, M|A, G]$; (2) $[C \otimes G, R|A, M]$; (3) $[R \otimes C|A, G, M]$; (4) $[C \otimes G|A, M, R]$; and (5) $[M \otimes R|A, C, G]$, respectively. Note that the latter three can be derived from the first two.

Collapsibility. If one is primarily interested in the relationships among the substance use variables, A, C, and M, then the best strategy would be to choose S $= \{A, G\}$ (see the graph below). Then Race can be collapsed over without affecting the first-order interactions λ^{AC}, λ^{AM}, or λ^{CM}.

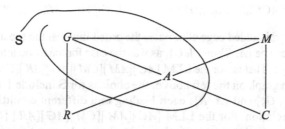

Alternatively, if one is primarily interested in how Gender relates to Alcohol and Marijuana Use (note that Gender is conditionally independent of Cigarette Use since λ^{GC} is absent from the LLM), then S should be chosen to be $\{A, M\}$ (see the graph below). Then Cigarette Use can be collapsed over without affecting the first-order interactions λ^{GA} and λ^{GM}.

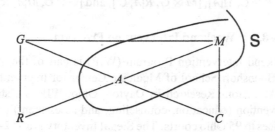

This example is a good illustration of the flexibility that the researcher has in choosing from the possible conditional independence relations so that the research goals and hypotheses are addressed in the most effective way. Similarly, there is flexibility in choosing which variables to collapse over depending on which associations are of most interest.

Note that the difference in the two competing LLMs for these data, $[AC][AM][CM][AG][AR][GR]$ and $[AC][AM][CM][AG][AR][GM][GR]$, can literally be seen in the association graphs; the dotted edge represents the added $[GM]$ term in the latter LLM:

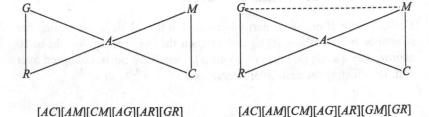

$[AC][AM][CM][AG][AR][GR]$ $[AC][AM][CM][AG][AR][GM][GR]$

Note that this added edge eliminates the possibility that the separating set of factors, S, can be $\{A, R\}$ or $\{A, C\}$, as was the case for the association graph on the left above. That is, for the LLM $[AC][AM][CM][AG][AR][GR]$, with the association graph on the left above, the choices for S include $\{A\}$, $\{A, M\}$, $\{A, C\}$, $\{A, G\}$, and $\{A, R\}$, each leading to a different conditional independence relation. For the LLM $[AC][AM][CM][AG][AR][GM][GR]$, with the association graph on the right above, the conditional independence relations $[C \otimes G, R|A, M]$ and $[R \otimes C, M|A, G]$ are obtained, with additional conditional independence relations derivable from these, as discussed above. By adding $[GM]$ to the generating class $[AC][AM][CM][AG][AR][GR]$, we've lost all instances in which G is conditionally independent of M; that is, we lose the conditional independencies $[G, R \otimes C, M|A]$, $[M \otimes G, R|A, C]$, and $[G \otimes C, M|A, R]$.

Example 8.2.3: Weekend Intervention Program

The Weekend Intervention Program (WIP) is part of the Wright State University Boonshoft School of Medicine Center for Interventions, Treatment, and Addictions Research in Dayton, Ohio. WIP is a state-certified driver intervention (education, counseling, and assessment) program providing services to 95 Ohio courts. The Siegal Inventory is a 152-item survey taken of WIP participants. Table 8.3, based on data for 3,599 males, provides the six factors used in this example. For notational purposes, the numerical labels corresponding to the factors (see Table 8.3) will be used in the following development, namely, 1 = Age, 2 = Marital Status, 3 = Number of Children, 4 = Education Level, 5 = Age at First Alcohol/Drug Use, and 6 = "Do you feel that you have a drinking problem?".

Table 8.3. Six Factors and Their Levels Taken From the Siegal Inventory Based on Data for 3,599 Males

Factor	Label	Levels
Age	1	$\leq 29, >29$
Marital Status	2	Never married, Other
Number of Children	3	$\geq 1, 0$
Education Level	4	$\leq$ High school, Other
Age at First Alcohol/Drug Use	5	≤ 14 years, >14 years
Do you feel that you have a drinking problem?	6	Yes, No

SOURCE: Data were kindly provided by Phyllis Cole, Director, Weekend Intervention Program, Wright State University Boonshoft School of Medicine, Dayton, Ohio.

NOTE: See Example 8.2.3.

Using a backward elimination scheme, an LLM that fits the data well is found to be the one having generating class [16][1345][1245][1234]. The association graph is given below:

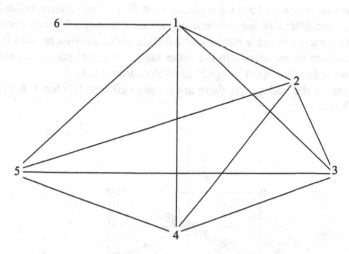

Decomposability. This LLM is not graphical because there is a maxclique, < 12345 >, appearing in the association graph, but there is no generator [12345] in the generating class. Hence, this model is not decomposable.

The multigraph, M, for [16][1345][1245][1234] is given below; a maximum spanning tree is shown in boldface:

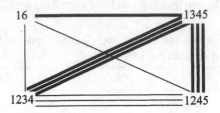

Using the calculation presented in Section 6.5, we have

Number of Indices Added Over Vertices	Number of Indices Added Over Branches	*Difference*	*Number of Factors*	*Decomposable?*
14	7	7	6	No

Just as we concluded from the association graph, this model is not decomposable.

FCIs and Interpretation. The only separation apparent in the association graph is between Factor 6 and the various subsets of Factors 2 through 5. Factor 1 must always be in the separation set, S. The implication is that Factor 6 is conditionally independent of Factors 2 through 5 given Factor 1. That is, for a given age, a participant's feeling about whether he has a drinking problem is not related to the other factors (marital status, number of children, education level, or age at first alcohol/drug use).

Based on the multigraph, there are six edge cutsets, labeled a, b, c, d, e, and f below:

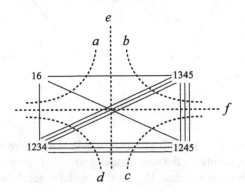

The edge cutsets and the indices involved are given below:

Edge Cutset	Indices
S_a	1
S_b	1, 3, 4, 5
S_c	1, 2, 4, 5
S_d	1, 2, 3, 4
S_e	1, 2, 3, 4
S_f	1, 3, 4, 5

The last two edge cutsets are duplicates of the preceding edge cutsets, so M/S multigraphs are provided for the first four edge cutsets:

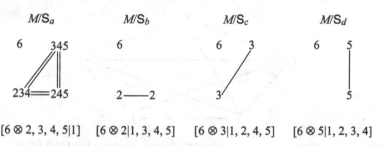

$$[6 \otimes 2, 3, 4, 5|1] \quad [6 \otimes 2|1, 3, 4, 5] \quad [6 \otimes 3|1, 2, 4, 5] \quad [6 \otimes 5|1, 2, 3, 4]$$

The latter three FCIs as well as all other conditional independencies can be derived from the first FCI above, $[6 \otimes 2, 3, 4, 5|1]$, by using the technique discussed in Section 6.7. In agreement with the conclusion derived from the association graph, Factor 6 is independent of Factors 2 through 5 given Factor 1. *Conclusion:* Regardless of age, the participant's feeling about whether he has a drinking problem is not strongly related to marital status, number of children, education level, or age at first alcohol/drug use.

Collapsibility. If one chooses, one may study the association between Factors 1 and 6 more closely by collapsing over the remaining factors (see the collapsibility conditions discussed in Sections 5.2 and 7.3).

Example 8.2.4: Occupational Aspirations

A classical data set involving occupational aspirations is provided in Agresti (2002; www.stat.ufl.edu/~aa/cda/cda.html). The variables and their levels are given in Table 8.4. Notationally, we have G = Gender, R = Residence, I = IQ, S = Socioeconomic Status, and O = Occupational Aspirations. One LLM that fits the data well is the one having generating class $[GRO][GSO][RSO][IO][IS][RI]$.

Table 8.4. Variables and Their Levels for the Occupational Aspirations
Data Set

Factor	Label	Levels
Gender	G	Male, Female
Residence	R	Rural, Small urban, Large urban
IQ	I	High, Low
Socioeconomic Status	S	High, Low
Occupational Aspirations	O	High, Low

SOURCE: Agresti (2002, p. 206; www.stat.ufl.edu/~aa/cda/cda.html).

NOTE: See Example 8.2.4.

The association graph is given below:

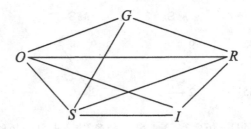

Decomposability. Note that $< GORS >$ is a maxclique but $[GORS]$ is not a
generator in the generating class. Therefore, this LLM is not decomposable
because it is not graphical.

The multigraph, M, for this LLM is given below, with a maximum spanning tree in boldface:

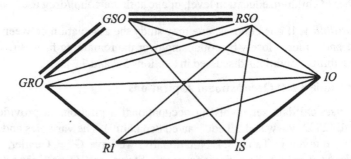

Given the calculation in the table below, it can be concluded that this
LLM is not decomposable.

Number of Indices Added Over Vertices	Number of Indices Added Over Branches	Difference	Number of Factors	Decomposable?
15	7	8	5	No

FCIs and Interpretation. The only missing edge preventing the association graph from being complete is the one joining *G* to *I*. Thus, the only possible choice for S is S $= \{O, R, S\}$ (see the graph below). That is, gender is independent of IQ conditional on residence, socioeconomic status, and occupational aspirations.

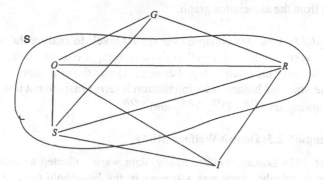

To find the conditional independencies from the multigraph, one must search for all the edge cutsets, S, and form *M*/S. In most cases, the resulting *M*/S multigraph has a single component, resulting in no conditional independence relation. The exception is the edge cutset given below (dotted line):

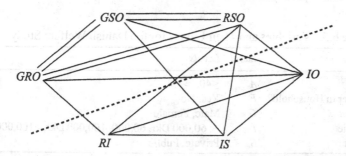

The indices contained in this edge cutset are *O*, *R*, and *S*. The multigraph *M*/S resulting from removing these indices is

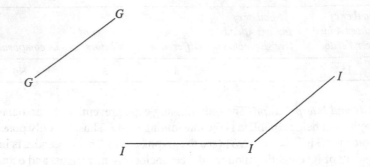

We conclude that $[G \otimes I | O, R, S]$, in agreement with the conclusion derived from the association graph.

Collapsibility. One may collapse over the IQ levels to analyze the relationships between Gender and the other three factors, or one may collapse over Gender to study the relationships between IQ and the remaining three factors. The first- and higher-order interaction λ-terms that are not invariant to the collapsing are λ^{OR}, λ^{OS}, λ^{RS}, and λ^{ORS}.

Example 8.2.5: Danish Welfare Study

In the 1976 Danish Welfare Study, data were collected to answer the question of whether there was a freezer in the household (see Andersen, 1997, pp. 124–126). The variables and their levels are given in Table 8.5. The variables are labeled as follows: A = Age, F = Freezer in Household, G = Gender, I = Income, and S = Sector. One of the LLMs that fits the data well has generating class $[GASF][GIF][IS]$. The multigraph approach will be used here since the multigraph is relatively small and easy to work with for this particular model.

Table 8.5. Variables and Their Levels for the Danish Welfare Study

Factor	Label	Levels
Age	A	≤ 40, > 40
Freezer in Household?	F	Yes, No
Gender	G	Male, Female
Income	I	< 60,000 Dkr, 60,000–100,000 Dkr, >100,000 Dkr
Sector	S	Private, Public

SOURCE: Data are given in Andersen (1997, p. 125).

NOTE: See Example 8.2.5.

The multigraph, *M*, with a maximum spanning tree in boldface, is shown below:

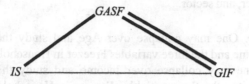

Decomposability. This model is not decomposable:

Number of Indices Added Over Vertices	Number of Indices Added Over Branches	*Difference*	*Number of Factors*	Decomposable?
9	3	6	5	No

FCIs and Interpretation. The three edge cutsets for this multigraph are shown below:

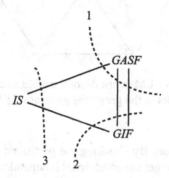

We have $S_1 = \{F, G, S\}$, $S_2 = \{F, G, I\}$, and $S_3 = \{I, S\}$. The multigraphs, $M/S_i, i = 1, 2, 3$ are

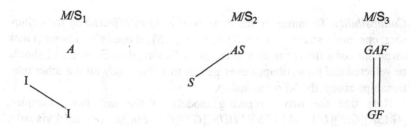

Only the first multigraph yields a conditional independence relation: $[A \otimes I|F, G, S]$. That is, age is independent of income given possession of a freezer, gender, and sector.

Collapsibility. One may collapse over Age and study the relationships between Income and the three variables Freezer in Household, Gender, and Sector. Or one may collapse over Income and study the relationships between Age and the three variables Freezer in Household, Gender, and Sector. The first- and higher-order λ-terms that are not invariant to collapsing are λ^{FG}, λ^{FS}, λ^{GS}, and λ^{FGS}.

Example 8.2.6: Danish Welfare Study, Alternative Model

Another model that fits these data well is the one having generating class $[AFS][FI][IS][GS][GI]$. This time, the association graph will be used:

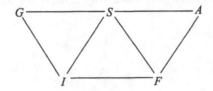

Decomposability. This LLM is not decomposable because though the max-clique $< FIS >$ appears in the graph the generator $[FIS]$ does not appear in the generating class.

FCIs and Interpretation. By choosing the partitions $S = \{F, I, S\}$, $T = \{G\}$, and $V = \{A\}$, we get the conditional independence relation $[G \otimes A|F, I, S]$. That is, gender is independent of age given possession of a freezer, income, and sector. By other choices of the partition, we get four additional conditional independence relations: $[G, I \otimes A|F, S]$; $[G \otimes A, F|I, S]$; $[A \otimes I|F, G, S]$; and $[F \otimes G|A, I, S]$.

Collapsibility. To minimize the number of λ-terms affected by the collapsing, one might select $S = \{F, S\}$ or $S = \{I, S\}$. If one is interested in how ownership of a freezer relates to the other factors, then $S = \{I, S\}$ should be selected and then collapse over gender to safely study all the other relationships except the *I-S* relationship.

Note that the two competing models of the last two examples, $[GASF][GIF][IS]$ and $[AFS][FI][IS][GS][GI]$, can be compared visually.

The association graphs are given below (the dotted edges represent the first-order interaction terms added to the second model to produce the first model):

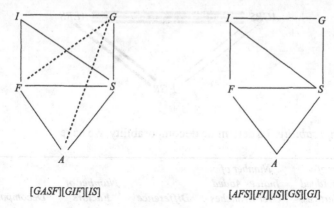

[GASF][GIF][IS]   [AFS][FI][IS][GS][GI]

The addition of the $<FG>$ and $<AG>$ edges to the association graph on the right above leads to the loss of any relationship in which F or A are conditionally independent of G. Thus, the five conditional independencies found in Example 8.2.6 for [AFS][FI][IS][GS][GI] are reduced to a single conditional independence relation, [$A \otimes I|F, G, S$], in Example 8.2.5 for [GASF][GIF][IS].

Example 8.2.7: Job Satisfaction Survey

A survey of a large national corporation is used to determine how Job Satisfaction (S) depends on Ethnic Group (E), Gender (G), Age (A), and Regional Location (R) (see Fowlkes, Freeny, & Landwehr, 1988). The variables and their levels are given in Table 8.6. One model that fits the data well has generating class [AGRS][AERS][EGR].

Table 8.6. Variables and Their Levels for the Job Satisfaction Survey

Factor	Label	Levels
Regional Location	R	Northeast, Mid-Atlantic, Southern, Midwest, Northwest, Southwest, Pacific
Ethnic Group	E	White, Other
Age	A	$< 35, 35$–$44, > 44$
Gender	G	Male, Female
Satisfied?	S	Yes, No

SOURCE: Fowlkes et al. (1988). Data can be found at www.stat.ufl.edu/~aa/cda/cda.html.

NOTE: See Example 8.2.7.

The multigraph, M, and a maximum spanning tree are given below:

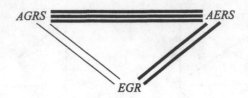

Decomposability. To determine decomposability, we have

Number of Indices Added Over Vertices	Number of Indices Added Over Branches	**Difference**	**Number of Factors**	Decomposable?
11	5	6	5	No

This LLM is not decomposable.

FCIs and Interpretation. There are three edge cutsets, as follows:

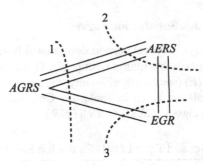

We have $S_1 = \{A, G, R, S\}$; $S_2 = \{A, E, R, S\}$; and $S_3 = \{E, G, R\}$.

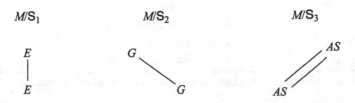

Based on the multigraphs M/S_i, $i = 1, 2$, and 3, there are no conditional independencies. This result is confirmed by the association graph, which is a complete graph. Consequently, there are no independence or conditional independence relationships for these data.

Collapsibility. Collapsing over any variable may change all the λ-terms in the LLM for the resulting marginal table.

Example 8.2.8: Marketing Rationale for Outsourcing Decisions

The strategic decision to outsource organizational functions (i.e., contract responsibility to a third party) is an important part of marketing strategy. This research compares and contrasts the marketing rationale of CEOs on their firm's outsourcing configuration. A sample of 384 firms from the Ewing Marion Kauffman Foundation's 1998 Survey of Innovative Practices (Cox & Camp, 1998) was used for the analysis by the Wright State University Statistical Consulting Center. The variables and their levels are given in Table 8.7. Notationally, we have O = Total Outsourcing Intensity, G = Geographic Scope, C = Control, T = Tenure, and F = Strategic Focus.

Table 8.7. Variables and Their Levels for the "Outsourcing Decisions" Data

Factor	Label	Levels
Total Outsourcing Intensity	O	None, 1–2, 3+
Geographic Scope	G	National, Local
Control	C	Managers, Partners, Control freaks
Tenure	T	Newbies, Founders
Strategic Focus	F	Prevention, Promotion

SOURCE: Data were kindly provided by Monte Shaffer, PhD student, Department of Marketing, Washington State University.

NOTE: See Example 8.2.8.

A backward elimination model selection procedure yields the LLM having generating class $[CTF][CFG][GO]$. The multigraph, with maximum spanning tree, for this model is given below:

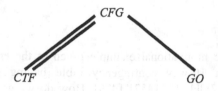

CFG

CTF   GO

Decomposability. This LLM is decomposable:

Number of Indices Added Over Vertices	Number of Indices Added Over Branches	Difference	Number of Factors	Decomposable?
8	3	5	5	Yes

FCIs and Interpretation. The conditional independencies can be derived directly from the branches of the maximum spanning tree by selecting S to be the branches. We have $S_1 = \{C, F\}$ and $S_2 = \{G\}$.

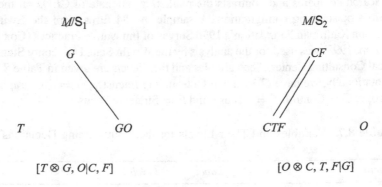

$$[T \otimes G, O|C, F] \qquad\qquad [O \otimes C, T, F|G]$$

We can draw the following conclusions:

1. Regardless of the CEO's level of control and strategic focus, tenure is independent of geographic scope and total outsourcing intensity.

2. Regardless of the geographic scope, total outsourcing intensity is independent of CEO control, tenure, and strategic focus.

Collapsibility. Depending on the researcher's interests, he or she may, among other possibilities, collapse over the levels of G and O to safely study the C-T-F, C-T, and T-F relationships in the three-way marginal table or collapse over C, F, and T to safely study the O-G relationship in the two-way marginal table.

8.3 Final Note

We return to the motivational example given at the end of Chapter 1, the LLM for a 10-way contingency table having generating class [67][013][125][178][1347][1457][1479]. How do we go about retrieving all the information (both methodological and practical) implied by this model in a reliable way? Even the most experienced LLM user might have trouble accomplishing this goal by scrutinizing the generating class or the LLM alone. Using the graphical procedures described by this book, the goal is accomplished reliably, methodically, and without using computer software, heavy calculations, or complex derivations, as follows:

- In Example 4.3.8, it is seen that the association graph for this LLM is rather complex, making information retrieval quite cumbersome, so we rely on the multigraph approach.

- The multigraph, given in Example 6.3.8, is also rather complex, but a maximum spanning tree is easily identified (Example 6.4.8).

- With a simple numerical calculation, it is determined that this model is decomposable (Example 6.5.8), hence it is both *graphical* (i.e., it can be interpreted exclusively in terms of independencies and conditional independencies without the need for interpreting high-order interactions) and *recursive* (the model can be oriented so as to have a causal interpretation). Furthermore, the FCIs are based exclusively on the branches of the maximum spanning tree, making them relatively easy to find.

- The joint distribution is obtained directly and explicitly from the vertices and branches of the maximum spanning tree (Example 6.6.8).

- The FCIs are obtained directly from the maximum spanning tree (Example 6.7.8). From the FCIs, all conditional independencies can be generated using the procedure given in Section 6.7.

- All possible collapsibility conditions can now be determined from the conditional independencies. For instance, in the first FCI of Example 6.7.8, $[0, 3 \otimes 2, 5 \otimes 6 \otimes 8 \otimes 9 | 1, 4, 7]$, Factors 0 and 3 are independent of Factors 2, 5, 6, 8, and 9 conditional on Factors 1, 4, and 7. So one may collapse over Factors 2, 5, 6, 8, and 9 and safely analyze the association between Factors 0 and 3 or between $\{0, 3\}$ and $\{1, 4, 7\}$, although the associations among Factors 1, 4, and 7 may not be preserved.

The two graphical methods presented in this book, the association graph and the multigraph, represent a marriage between statistics and mathematics. In particular, graph theory principles are applied to a statistical model, the LLM, for the purpose of analyzing the structural associations among a set of categorical variables. This application of graph theory to LLM theory is based on results deriving from mathematical subject areas such as engineering, computer science, linear algebra, and, of course, graph theory (see Khamis & McKee, 1997; McKee & Khamis, 1996).

In alluding to data mining methods, Agresti (2002, p. 631) predicts that more work will be done involving "huge data sets with large numbers of variables." The need for analyzing and interpreting large, complex LLMs is growing as researchers gain access to ever larger data sets. With this book, the researcher faced with this need is equipped with two alternative graphical procedures that accomplish the goal of analyzing and interpreting such models in a facile, organized, comprehensive manner.

The very first statement given in this book, "*Il faut toujours prendre la mesure des choses avant de décider*" ("It's always necessary to take the measure of things before deciding"), is one that all scientists can agree on. But perhaps even more important, it is necessary to accurately and reliably interpret those measures before arriving at conclusions. Use of the procedures in this book will inevitably lead to a better understanding of categorical variable relationships and, hence, to more accurate, comprehensive, and explicit conclusions concerning the data. In turn, this leads to better science, which is the ultimate goal of methodological research.

DATA SETS

*Catholic Survey of Support of
Women in the Priesthood*
Source: www.thearda.com/
Archive/Files/Codebooks/
GALLUP99_CB.asp
Example 3.3.1
Tables 3.3, 3.4

Danish Welfare Study
Source: Andersen (1997, pp.
124–126)
Examples 8.2.5, 8.2.6
Table 8.5

*Dayton High School Alcohol/
Drug Use Survey, conducted
collaboratively by the Wright
State University Boonshoft
School of Medicine and United
Health Services*
Source: Wright State University
Statistical Consulting Center.
Client: Dr. Russel Falk
Examples 2.2.1, 2.4.1, 4.3.7,
5.3.1, 6.3.7, 6.4.7, 6.5.7, 6.6.7,
6.7.7, 7.1.7, 7.2.7, 8.2.2
Tables 2.1, 2.2, 4.2, 5.6

Drug Treatment Refusal Study
Source: Wright State University
Statistical Consulting Center.
Client: Dr. Paul Rodenhauser
Example 5.1.1
Tables 5.1, 5.2

Florida Murder Cases
Source: Radelet and Pierce
(1991)
Example 5.1.2
Tables 5.3, 5.4

*Institute for Social Research,
Copenhagen Survey*
Source: Edwards and Kreiner
(1983)
Examples 4.3.6, 5.2.2, 6.3.6,
6.4.6, 6.5.6, 6.6.6, 6.7.6, 7.1.6,
7.2.6
Table 4.1

Job Satisfaction Survey
Source: Fowlkes et al. (1988)
Example 8.2.7
Table 8.6

*Marketing Rationale for
Outsourcing Decisions*
Source: Wright State University
Statistical Consulting Center.
Client: Mr. Monte Shaffer
(unpublished data)
Example 8.2.8
Table 8.7

Occupational Aspirations
Source: Agresti (2002; www.stat
.ufl.edu/~aa/cda/cda.html)
Example 8.2.4
Table 8.4

114

PASS Program for Student Retention
Source: Wright State University Statistical Consulting Center.
Client: Dr. Anita Curry-Jackson
Example 8.2.1
Table 8.2

Weekend Intervention Program
Source: Wright State University Statistical Consulting Center.
Client: Ms. Phyllis Cole
Example 8.2.3
Table 8.3

REFERENCES

Agresti, A. (1984). *Analysis of ordinal categorical data*. New York: Wiley.

Agresti, A. (2002). *Categorical data analysis* (2nd ed.). New York: Wiley-Interscience.

Andersen, A. H. (1974). Multidimensional contingency tables. *Scandinavian Journal of Statistics, 1,* 115–127.

Andersen, E. B. (1997). *Introduction to the statistical analysis of categorical data*. New York: Springer.

Asmussen, S., & Edwards, D. (1983). Collapsibility and response variables in contingency tables. *Biometrika, 70,* 567–578.

Bartlett, M. S. (1935). Contingency table interactions. *Journal of the Royal Statistical Society, Suppl. 2,* 248–252.

Benedetti, J. K., & Brown, M. B. (1978). Strategies for the selection of loglinear models. *Biometrics, 34,* 680–686.

Birch, M. W. (1963). Maximum likelihood in three-way contingency tables. *Journal of the Royal Statistical Society, Series B, 25,* 220–233.

Birch, M. W. (1965). The detection of partial association II: The general case. *Journal of the Royal Statistical Society, Series B, 27,* 111–124.

Bishop, Y. M. M. (1971). Effects of collapsing multidimensional contingency tables. *Biometrics, 27,* 545–562.

Bishop, Y. M. M., Fienberg, S. E., & Holland, P. W. (1975). *Discrete multivariate analysis*. Cambridge: MIT Press.

Blair, P. A., & Peyton, B. W. (1993). An introduction to chordal graphs and clique trees. In J. A. George, J. R. Gilbert, & J. W. H. Liu (Eds.), *Graph theory and sparse matrix computations* (pp. 1–10). IMA Volumes in Mathematics and Its Applications, No. 56. Berlin, Germany: Springer.

Brown, M. B. (1976). Screening effects in multidimensional contingency tables. *Applied Statistics, 25,* 37–46.

Christensen, R. (1990). *Log-linear models*. New York: Springer.

Cochran, W. G. (1954). Some methods for strengthening the common χ^2 tests. *Biometrics, 10,* 417–451.

Cox, L. W., & Camp, S. M. (1998). *Survey of innovative practices: 1999 executive report* (Tech. Rep.). Kansas City, MO: Kauffman Center for Entrepreneurial Leadership.

Darroch, J. N., Lauritzen, S. L., & Speed, T. P. (1980). Markov fields and log-linear interaction models for contingency tables. *Annals of Statistics, 8,* 522–539.

Edwards, D. (1995). *Introduction to graphical modelling*. New York: Springer.

115

Edwards, D., & Kreiner, S. (1983). The analysis of contingency tables by graphical methods. *Biometrika, 70,* 553–565.

Fienberg, S. E. (1979). The use of chi-squared statistics for categorical data problems. *Journal of the Royal Statistical Society, Series B, 41,* 54–64.

Fienberg, S. E. (1981). *The analysis of cross-classified categorical data* (2nd ed.). Cambridge: MIT Press.

Fisher, R. A. (1925). *Statistical methods for research workers.* Edinburgh, UK: Oliver & Boyd.

Fleiss, J., Levin, B., & Paik, M. C. (2003). *Statistical methods for rates and proportions* (3rd ed.). Hoboken, NJ: Wiley.

Fowlkes, E. B., Freeny, A. E., & Landwehr, J. (1988). Evaluating logistic models for large contingency tables. *Journal of the American Statistical Association, 83,* 611–622.

Gibbons, A. (1985). *Algorithmic graph theory.* Cambridge, UK: Cambridge University Press.

Golumbic, M. C. (1980). *Algorithmic graph theory and perfect graphs.* San Diego, CA: Academic Press.

Good, I. J., & Mittal, Y. (1987). The amalgamation and geometry of two-by-two contingency tables. *Annals of Statistics, 15,* 694–711.

Goodman, L. A. (1970). The multivariate analysis of qualitative data: Interaction among multiple classifications. *Journal of the American Statistical Association, 65,* 226–256.

Goodman, L. A. (1971a). The analysis of multidimensional contingency tables: Stepwise procedures and direct estimation methods for building models for multiple classifications. *Technometrics, 13,* 33–61.

Goodman, L. A. (1971b). Partitioning of chi-square, analysis of marginal contingency tables, and estimation of expected frequencies in multidimensional contingency tables. *Journal of the American Statistical Association, 66,* 339–344.

Goodman, L. A. (1973). The analysis of contingency tables when some variables are posterior to others: A modified path analysis approach. *Biometrika, 60,* 179–192.

Goodman, L. A. (2007). Statistical magic and/or statistical serendipity: An age of progress in the analysis of categorical data. *Annual Review of Sociology, 33,* 1–19.

Goodman, L. A., & Kruskal, W. H. (1979). *Measures of association for cross classifications.* New York: Springer.

Grizzle, J. E., Starmer, C. F., & Koch, G. G. (1969). Analysis of categorical data by linear models. *Biometrics, 25,* 489–504.

Haberman, S. J. (1974). *The analysis of frequency data* (IMS Monographs). Chicago: University of Chicago Press.

Khamis, H. J. (1983). Log-linear model analysis of the semi-symmetric intraclass contingency table. *Communications in Statistics Series A, 12,* 2723–2752.

Khamis, H. J. (1996). Application of the multigraph representation of hierarchical log-linear models. In A. von Eye & C. C. Clogg (Eds.), *Categorical variables in developmental research: Methods of analysis* (pp. 215–232). New York: Academic Press.

Khamis, H. J. (2004). Measures of association. In P. Armitage & T. Colton (Eds.), *Encyclopedia of biostatistics* (pp. 236–241). New York: Wiley.

Khamis, H. J. (2005). Multigraph modeling. In B. Everitt & D. Howell (Eds.), *Encyclopedia of statistics in behavioral science* (pp. 1294–1296). New York: Wiley.

Khamis, H. J., & McKee, T. A. (1997). Chordal graph models of contingency tables. *Computers and Mathematics With Applications, 34,* 89–97.

Knoke, D., & Burke, P. B. (1980). *Log-linear models.* Sage University Papers Series on Quantitative Applications in the Social Sciences, No. 07-020. Beverly Hills, CA: Sage.

Koehler, K. J. (1986). Goodness-of-fit tests for loglinear models in sparse contingency tables. *Journal of the American Statistical Association, 81,* 483–493.

Kruskal, J. B. (1956). On the shortest spanning sub-tree and the travelling salesman problem. *Proceedings of the American Mathematical Society, 7,* 48–50.

Lauritzen, S. L. (1996). *Graphical models.* Oxford, UK: Clarendon Press.

Lawal, B. (2003). Categorical data analysis with SAS and SPSS applications. Mahwah, NJ: Lawrence Erlbaum.

Lawal, H. B., & Upton, G. J. G. (1984). On the use of χ^2 as a test of independence in contingency tables with small cell expectations. *Australian Journal of Statistics, 26,* 75–85.

Lee, S. K. (1977). On the asymptotic variance of $\hat{u}$-terms in loglinear models of multidimensional contingency tables. *Journal of the American Statistical Association, 72,* 412–419.

McKee, T. A., & Khamis, H. J. (1996). Multigraph representations of hierarchical loglinear models. *Journal of Statistical Planning and Inference, 53,* 63–74.

Pearl, J. (1988). *Probabilistic reasoning in intelligent systems: Networks of plausible inference.* San Mateo, CA: Morgan Kaufman.

Radelet, M. L., & Pierce, G. L. (1991). Choosing those who will die: Race and the death penalty in Florida. *Florida Law Review, 43,* 1–34.

Rodenhauser, P., Schwenkner, C., & Khamis, H. (1987). Factors related to drug treatment refusal in a forensic hospital. *Hospital and Community Psychiatry, 38,* 631–637.

Roscoe, J. T., & Byars, J. A. (1971). Sample size restraints commonly imposed on the use of the chi-square statistic. *Journal of the American Statistical Association, 66,* 755–759.

Rudas, T. (1998). *Odds ratios in the analysis of contingency tables.* Sage University Papers Series on Quantitative Applications in the Social Sciences, No. 07-119. Thousand Oaks, CA: Sage.

Savant, M. vos. (1996, April 28). One company's hiring experience: Did it discriminate without knowing it? *Parade Magazine,* 6–7.

Simpson, E. H. (1951). The interpretation of interaction in contingency tables. *Journal of the Royal Statistical Society, Series B, 13,* 238–241.

Stewart, R. D., Paris, P. M., Pelton, G. H., & Garretson, D. (1984). Effect of varied training techniques on field endotracheal intubation success rates. *Annals of Emergency Medicine, 13,* 1032–1036.

118

Tarjan, R. E., & Yannakakis, M. (1984). Simple linear-time algorithms to test chordality of graphs, test acyclicity of hypergraphs, and selectively reduce acyclic hypergraphs. *SIAM Journal on Computing, 13,* 566–579.

Wagner, C. H. (1982). Simpson's paradox in real life. *The American Statistician, 36,* 46–48.

Wermuth, N. (1980). Linear recursive equations, covariance selection, and path analysis. *Journal of the American Statistical Association, 75,* 963–972.

Wermuth, N., & Lauritzen, S. L. (1983). Graphical and recursive models for contingency tables. *Biometrika, 70,* 537–552.

Whittaker, J. (1990). *Graphical models in applied multivariate statistics.* New York: Wiley.

Wickens, T. D. (1989). *Multiway contingency tables analysis for the social sciences.* Hillsdale, NJ: Lawrence Erlbaum.

Wilson, R. J. (1985). *Introduction to graph theory* (3rd ed.). Harlow, UK: Longman.

AUTHOR INDEX

119

SUBJECT INDEX

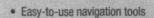

Printed in the United States
By Bookmasters